Two Manuals Combined into One Book

Special Forces Survival Evasion and Recovery

And

Soldier's Handbook for Individual Operations Survival in Cold Weather Areas

Department of Defense

Pentagon
Publishing

QUICK REFERENCE CHECKLIST

Decide to Survive!

S -	**S**ize up the situation, surroundings, physical condition, equipment.
U -	**U**se all your senses
R -	**R**emember where you are.
V -	**V**anquish fear and panic.
I -	**I**mprovise and improve.
V -	**V**alue living.
A -	**A**ct like the natives.
L -	**L**ive by your wits.

1. Immediate Actions

a. Assess immediate situation. ***THINK BEFORE YOU ACT!***

b. Take action to protect yourself from nuclear, biological, or chemical hazards (Chapter IX).

c. Seek a concealed site.

d. Assess medical condition; treat as necessary (Chapter V).

e. Sanitize uniform of potentially compromising information.

f. Sanitize area; hide equipment you are leaving.

g. Apply personal camouflage.

h. Move away from concealed site, zigzag pattern recommended.

i. Use terrain to advantage, communication, and concealment.

j. Find a hole-up site.

2. Hole-Up-Site (Chapter I)

a. Reassess situation; treat injuries, then inventory equipment.

b. Review plan of action; establish priorities (Chapter VI).

c. Determine current location.

d. Improve camouflage.

e. Focus thoughts on task(s) at hand.

f. Execute plan of action. Stay flexible!

Recommend inclusion of this manual in the aviator's survival vest.

3. Concealment (Chapter I)

a. Select a place of concealment providing—

(1) Adequate concealment, ground and air.

(2) Safe distance from enemy positions and lines of communications (LOC).

(3) Listening and observation points.

(4) Multiple avenues of escape.

(5) Protection from the environment.

(6) Possible communications/signaling opportunities.

b. Stay alert, maintain security.

c. Drink water.

4. Movement (Chapters I and II)

a. Travel slowly and deliberately.

b. ***DO NOT*** leave evidence of travel; use noise and light discipline.

c. Stay away from LOC.

d. Stop, look, listen, and smell; take appropriate action(s).

e. Move from one concealed area to another.

f. Use evasion movement techniques (Chapter I).

5. Communications and Signaling (Chapter III)

a. Communicate as directed in applicable plans/orders, particularly when considering transmitting ***in the blind***.

b. Be prepared to use communications and signaling devices on short notice.

c. Use of communications and signaling devices may compromise position.

6. Recovery (Chapter IV)

a. Select site(s) IAW criteria in theater recovery plans.

b. Ensure site is free of hazards; secure personal gear.

c. Select best area for communications and signaling devices.

d. Observe site for proximity to enemy activity and LOC.

e. Follow recovery force instructions.

FM 21-76-1
MCRP 3-02H
NWP 3-50.3
AFTTP(I) 3-2.26

FM 21-76-1 — U.S. Army Training and Doctrine Command, Fort Monroe, Virginia
MCRP 3-02H — Marine Corps Combat Development Command, Quantico, Virginia
NWP 3-50.3 — Navy Warfare Development Command, Newport, Rhode Island
AFTTP(I) 3-2.26 — Headquarters Air Force Doctrine Center, Maxwell Air Force Base, Alabama

29 JUNE 1999

Survival, Evasion, and Recovery

Multiservice Procedures for Survival, Evasion, and Recovery

Note: This *UNCLASSIFIED* publication is designed to provide Service members quick-reference survival, evasion, and recovery information. See Appendix B for the scope, purpose, application, implementation plan, and user information.

TABLE OF CONTENTS

THE CODE OF CONDUCT

ARTICLE I

I am an American, fighting in the forces which guard my country and our way of life. I am prepared to give my life in their defense.

ARTICLE II

I will never surrender of my own free will. If in command, I will never surrender the members of my command while they still have the means to resist.

ARTICLE III

If I am captured, I will continue to resist by all means available. I will make every effort to escape and aid others to escape. I will accept neither parole nor special favors from the enemy.

ARTICLE IV

If I become a prisoner of war, I will keep faith with my fellow prisoners. I will give no information or take part in any action which might be harmful to my comrades. If I am senior, I will take command. If not, I will obey the lawful orders of those appointed over me and will back them up in every way.

ARTICLE V

When questioned, should I become a prisoner of war, I am required to give name, rank, service number and date of birth. I will evade answering further questions to the utmost of my ability. I will make no oral or written statements disloyal to my country and its allies or harmful to their cause.

ARTICLE VI

I will never forget that I am an American, fighting for freedom, responsible for my actions, and dedicated to the principles which made my country free. I will trust in my God and in the United States of America.

Chapter I

EVASION

1. Planning

a. Review the quick reference checklist on the inside cover.

b. Guidelines for successful evasion include—

(1) Keeping a positive attitude.

(2) Using established procedures.

(3) Following your evasion plan of action.

(4) Being patient.

(5) Drinking water (***DO NOT*** eat food without water).

(6) Conserving strength for critical periods.

(7) Resting and sleeping as much as possible.

(8) Staying out of sight.

c. The following odors stand out and may give an evader away:

(1) Scented soaps and shampoos.

(2) Shaving cream, after-shave lotion, or other cosmetics.

(3) Insect repellent (camouflage stick is least scented).

(4) Gum and candy (smell is strong or sweet).

(5) Tobacco (odor is unmistakable).

d. Where to go (initiate evasion plan of action):

(1) Near a suitable area for recovery.

(2) Selected area for evasion.

(3) Neutral or friendly country or area.

(4) Designated area for recovery.

2. Camouflage

a. Basic principles:

(1) Disturb the area as little as possible.

(2) Avoid activity that reveals movement to the enemy.

(3) Apply personal camouflage.

b. Camouflage patterns **(Figure I-1)**:

(1) Blotch pattern.

(a) Temperate deciduous (leaf shedding) areas.

(b) Desert areas (barren).

(c) Snow (barren).

(2) Slash pattern.

(a) Coniferous areas (broad slashes).

(b) Jungle areas (broad slashes).
(c) Grass (narrow slashes).
(3) Combination. May use blotched and slash together.

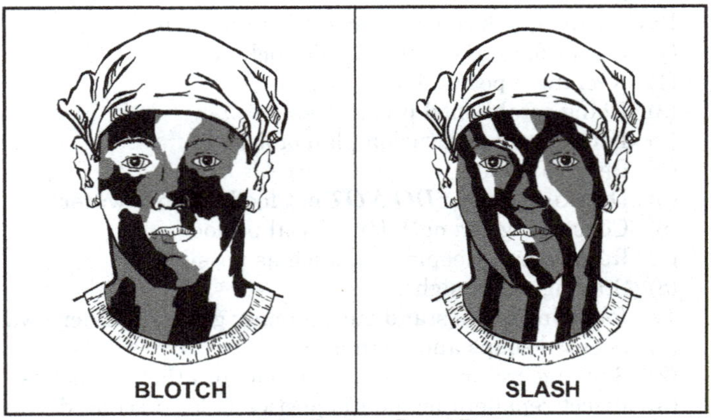

Figure I-1. Camouflage Patterns

c. Personal camouflage application follows:

(1) Face. Use dark colors on high spots and light colors on any remaining exposed areas. Use a hat, netting, or mask if available.

(2) Ears. The insides and the backs should have **2** colors to break up outlines.

(3) Head, neck, hands, and the under chin. Use scarf, collar, vegetation, netting, or coloration methods.

(4) Light colored hair. Give special attention to conceal with a scarf or mosquito head net.

d. Position and movement camouflage follows:

(1) Avoid unnecessary movement.
(2) Take advantage of natural concealment:
(a) Cut foliage fades and wilts, change regularly.
(b) Change camouflage depending on the surroundings.
(c) ***DO NOT*** select vegetation from same source.
(d) Use stains from grasses, berries, dirt, and charcoal.
(3) ***DO NOT*** over camouflage.
(4) Remember when using shadows, they shift with the sun.

(5) Never expose shiny objects (like a watch, glasses, or pens).

(6) Ensure watch alarms and hourly chimes are turned off.

(7) Remove unit patches, name tags, rank insignia, etc.

(8) Break up the outline of the body, ***"V"*** of crotch/armpits.

(9) Conduct observation from a prone and concealed position.

3. Shelters

a. Use camouflage and concealment.

b. Locate carefully—easy to remember acronym: ***BLISS***.

B - **B**lend **L** - **L**ow silhouette **I** - **I**rregular shape **S** - **S**mall **S** - **S**ecluded location

(1) Choose an area—

(a) Least likely to be searched (drainages, rough terrain, etc.) and blends with the environment.

(b) With escape routes (***DO NOT*** corner yourself).

(c) With observable approaches.

(2) Locate entrances and exits in brush and along ridges, ditches, and rocks to keep from forming paths to site.

(3) Be wary of flash floods in ravines and canyons.

(4) Conceal with minimal to no preparation.

(5) Take the direction finding threat into account before transmitting from shelter.

(6) Ensure overhead concealment.

4. Movement

a. A moving object is easy to spot. If travel is necessary—

(1) Mask with natural cover **(Figure I-2)**.

(2) Use the military crest.

(3) Restrict to periods of low light, bad weather, wind, or reduced enemy activity.

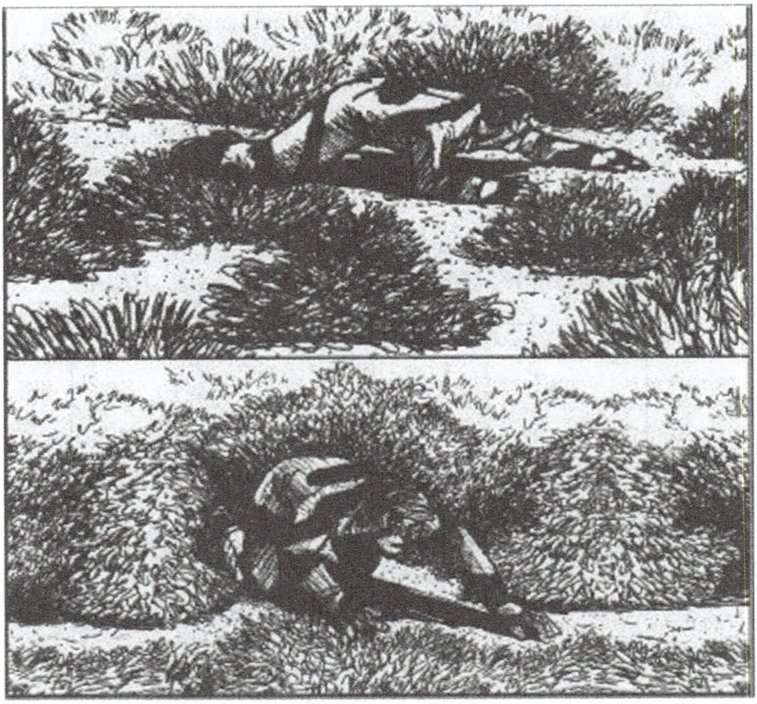

Figure I-2. Ground Movement

(4) Avoid silhouetting **(Figure I-3)**.

(5) At irregular intervals—

(a) ***STOP*** at a point of concealment.

(b) ***LOOK*** for signs of human or animal activity (smoke, tracks, roads, troops, vehicles, aircraft, wire, buildings, etc.). Watch for trip wires or booby traps and avoid leaving evidence of travel. Peripheral vision is more effective for recognizing movement at night and twilight.

(c) ***LISTEN*** for vehicles, troops, aircraft, weapons, animals, etc.

(d) ***SMELL*** for vehicles, troops, animals, fires, etc.

Figure I-3. Avoid Silhouetting

(6) Employ noise discipline; check clothing and equipment for items that could make noise during movement and secure them.

b. Break up the human shape or recognizable lines.

c. Route selection requires detailed planning and special techniques (irregular route/zigzag) to camouflage evidence of travel.

d. Some techniques for concealing evidence of travel follows:

(1) Avoid disturbing the vegetation above knee level.

(2) ***DO NOT*** break branches, leaves, or grass.

(3) Use a walking stick to part vegetation and push it back to its original position.

(4) ***DO NOT*** grab small trees or brush. (This may scuff the bark or create movement that is easily spotted. In snow country, this creates a path of snowless vegetation revealing your route.)

(5) Pick firm footing (carefully place the foot lightly but squarely on the surface to avoid slipping). ***TRY NOT TO***—

(a) Overturn ground cover, rocks, and sticks.

(b) Scuff bark on logs and sticks.

(c) Make noise by breaking sticks. (Cloth wrapped around feet helps muffle this.)

(d) Mangle grass and bushes that normally spring back.

(6) Mask unavoidable tracks in soft footing by—

(a) Placing tracks in the shadows of vegetation, downed logs, and snowdrifts.

(b) Moving before and during precipitation allows tracks to fill in.

(c) Traveling during windy periods.

(d) Taking advantage of solid surfaces (logs, rocks, etc.) leaving less evidence of travel.

(e) Patting out tracks lightly to speed their breakdown or make them look old.

(7) Secure trash or loose equipment—hide or bury discarded items. (Trash or lost equipment identifies who lost it.)

(8) Concentrate on defeating the handler if pursued by dogs.

e. Penetrate obstacles as follows:

(1) Enter deep ditches feet first to avoid injury.

(2) Go around chain-link and wire fences. Go under fence if unavoidable, crossing at damaged areas. ***DO NOT*** touch fence; look for electrical insulators or security devices.

(3) Penetrate rail fences, passing under or between lower rails. If impractical, go over the top, presenting as low a silhouette as possible **(Figure I-4)**.

(4) Cross roads after observation from concealment to determine enemy activity. Cross at points offering concealment such as bushes, shadows, bend in road, etc. Cross in a manner leaving your footprints parallel (cross step sideways) to the road. **(Figure I-5)**

(5) Use same method of observation for railroad tracks that was used for roads. Next, align body parallel to tracks with face down, cross tracks using a semi-pushup motion. Repeat for the second track. **(Figure I-6)**.

Figure I-4. Rail Fences

Figure I-5. Road Crossing

Figure I-6. Railroad Tracks

WARNING: If 3 rails exist, 1 may be electrified.

Chapter II
NAVIGATION

Assess the threat and apply appropriate evasion principles.

1. Stay or Move Considerations

a. Stay with the vehicle/aircraft in a non-combat environment.

b. Leave only when—

(1) Dictated by the threat.

(2) Are certain of your location, have a known destination, and have the ability to get there.

(3) Can reach water, food, shelter, and/or help.

(4) Convinced rescue is not coming.

c. Consider the following if you decide to travel:

(1) Follow the briefed evasion plan.

(2) Determine which direction to travel and why.

(3) Decide what equipment to take, cache, or destroy.

d. Leave information at your starting point (in a non-combat environment) that includes—

(1) Destination.

(2) Route of travel.

(3) Personal condition.

(4) Supplies available.

e. Consider the following for maps (in a combat environment):

(1) ***DO NOT*** write on the map.

(2) ***DO NOT*** soil the map by touching the destination.

(3) ***DO NOT*** fold in a manner providing travel information.

Note: **These actions may compromise information if captured.**

2. Navigation and Position Determination

a. Determine your general location by—

(1) Developing a working knowledge of the operational area.

(a) Geographic checkpoints.

(b) Man-made checkpoints.

(c) Previous knowledge of operational area.

(2) Using the ***Rate x Time = Distance*** formula.

(3) Using information provided in the map legend.

(4) Using prominent landmarks.

(5) Visualizing map to determine position.

b. Determine cardinal directions (north, south, east, and west) by—

(1) Using compass.

CAUTION: The following methods are ***NOT*** highly accurate and give only general cardinal direction.

(2) Using stick and shadow method to determine a true north-south line **(Figure II-1)**.

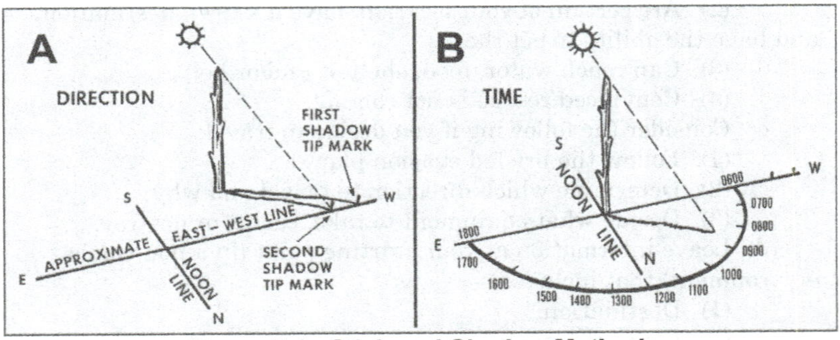

Figure II-1. Stick and Shadow Method

(3) Remembering the sunrise/moonrise is in the east and sunset/moonset is in the west.

(4) Using a wristwatch to determine general cardinal direction **(Figure II-2)**.

(a) Digital watches. Visualize a clock face on the watch.

(b) Northern Hemisphere. Point hour hand at the sun. South is halfway between the hour hand and 12 o'clock position.

(c) Southern Hemisphere. Point the 12 o'clock position on your watch at the sun. North is halfway between the 12 o'clock position and the hour hand.

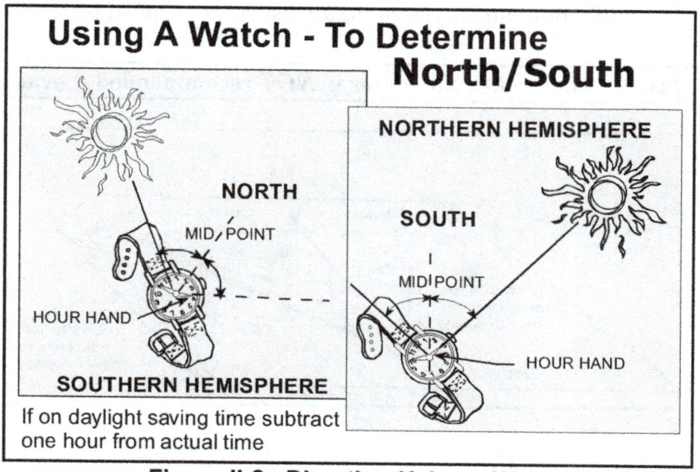

Figure II-2. Direction Using a Watch

(5) Using a pocket navigator **(Figure II-3)**—

(a) Gather the following necessary materials:

•Flat writing material (such as an MRE box).

•1-2 inch shadow tip device (a twig, nail, or match).

•Pen or pencil.

(b) Start construction at sunup; end construction at sundown. Do the following:

•Attach shadow tip device in center of paper.

•Secure navigator on flat surface (***DO NOT*** move during set up period).

•Mark tip of shadow every 30 minutes annotating the time.

•Connect marks to form an arc.

•Indicate north with a drawn arrow.

Note: The shortest line between base of shadow tip device and curved line is a north-south line.

(c) Do the following during travel:

•Hold navigator so the shadow aligns with mark of present time (drawn arrow now points to true north).

(d) Remember the navigator is current for approximately 1 week.

CAUTION: The Pocket Navigator is ***NOT*** recommended if evading.

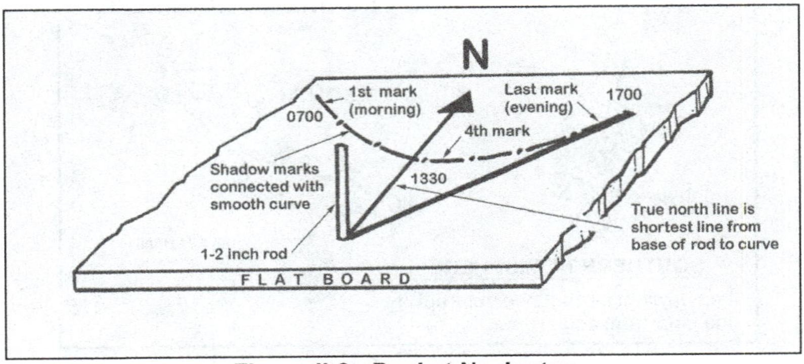

Figure II-3. Pocket Navigator

(6) Using the stars **(Figure II-4)** the—

(a) North Star is used to locate true north-south line.

(b) Southern Cross is used to locate true south-north line.

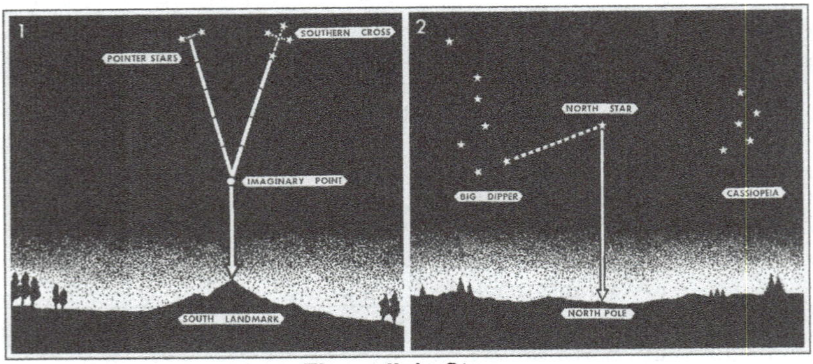

Figure II-4. Stars

c. Orient the map by—

(1) Using a true north-south line **(Figure II-5)**—

(a) Unfold map and place on a firm, flat, level nonmetallic surface.

(b) Align the compass on a true north-south line.

(c) Rotate map and compass until stationary index line aligns with the magnetic variation indicated in marginal information.

- Easterly (subtract variation from 360 degrees).
- Westerly (add variation to 360 degrees).

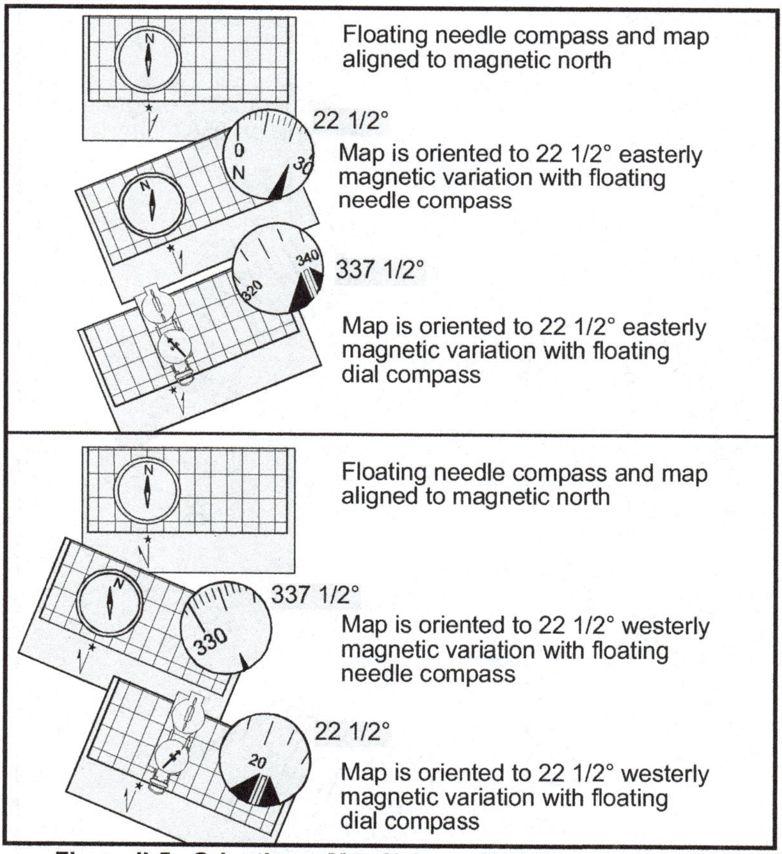

Figure II-5. Orienting a Map Using a True North-South Line

(2) Using a compass rose **(Figure II-6)**—

(a) Place edge of the lensatic compass on magnetic north line of the compass rose closest to your location.

(b) Rotate map and compass until compass reads 360 degrees.

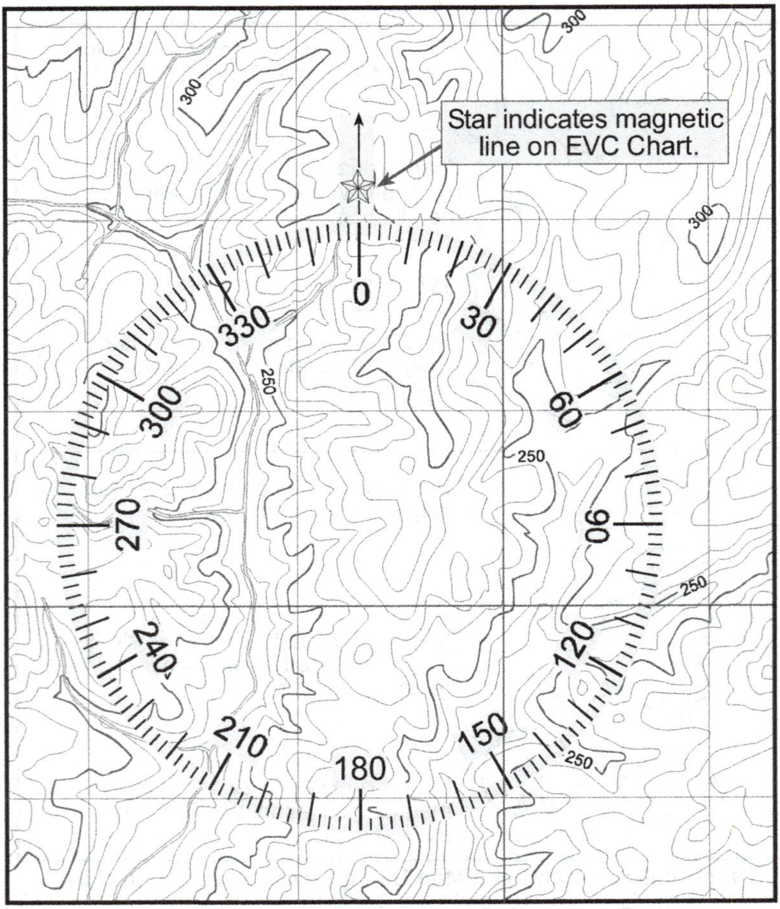

Figure II-6. Map Orientation with Compass Rose

(3) If there is **NO** compass, orient map using cardinal direction obtained by the stick and shadow method or the celestial aids (stars) method.

d. Determine specific location.

(1) Global Positioning System (GPS).

(a) ***DO NOT*** use GPS for primary navigation.

(b) Use GPS to confirm your position ***ONLY***.

(c) Select area providing maximum satellite reception.

(d) Conserve GPS battery life.

(2) Triangulation (resection) with a compass **(Figure II-7)**.

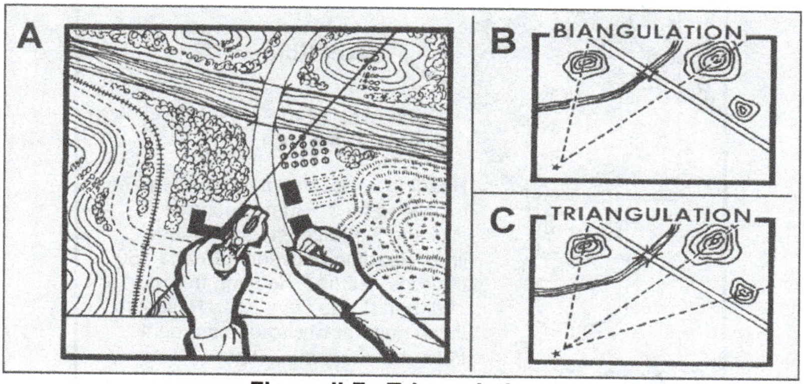

Figure II-7. Triangulation

(a) Try to use **3** or more azimuths.

(b) Positively identify a major land feature and determine a line of position (LOP).

(c) Check map orientation each time compass is used.

(d) Plot the LOP using a thin stick or blade of grass (combat) or pencil line (non-combat).

(e) Repeat steps **(b)** through **(d)** for other LOPs.

e. Use the compass for night navigation by—

(1) Setting up compass for night navigation **(Figure II-8)**.

(2) Aligning north-seeking arrow with luminous line and follow front of compass.

(3) Using point-to-point navigation.

f. Route selection techniques follow:

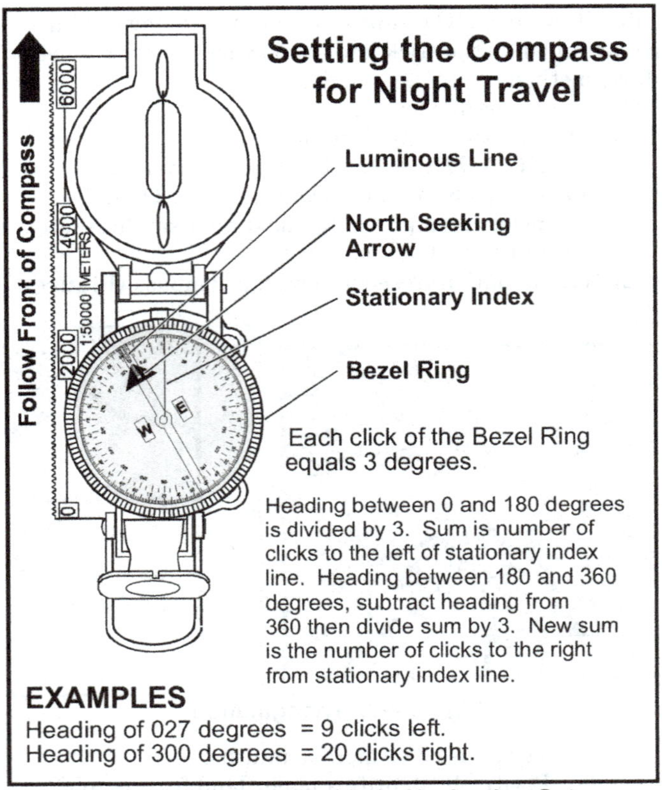

Figure II-8. Compass Night Navigation Setup

(1) Circumnavigation.

(a) Find a prominent landmark on the opposite side of the obstacle.

(b) Contour around obstacle to landmark.

(c) Resume your route of travel.

(2) Dogleg and 90 degree offset **(Figure II-9)**.

(3) Straight-line heading as follows:

(a) Maintain heading until reaching destination.

(b) Measure distance by counting the number of paces in a given course and convert to map units.

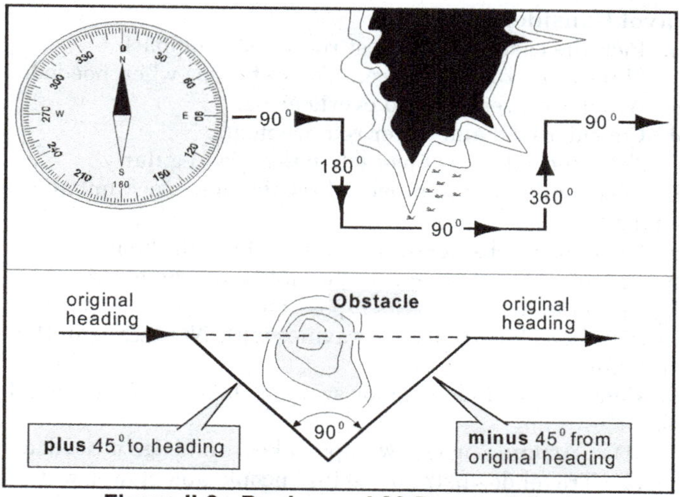

Figure II-9. Dogleg and 90 Degree Offset

•One pace is the distance covered each time the same foot touches the ground.

•Distances measured by paces are approximate (example in open terrain, 900 paces per kilometer [average], or example in rough terrain, 1200 paces per kilometer [average]).

(c) Use pace count in conjunction with terrain evaluation and heading to determine location. An individual's pace varies because of factors such as steep terrain, day/night travel, or injured/uninjured condition. Adjust estimation of distance traveled against these factors to get relative accuracy when using a pace count.

(4) Deliberate offset is—

(a) Used when finding a point on a linear feature (that is, road or river).

(b) Intentionally navigated to left or right of target so you know which way to turn at the linear feature.

(5) Point-to-point is same as straight line.

(a) Pick out landmarks on the heading and walk the trail of least resistance to a point.

(b) On reaching a point, establish another landmark and continue.

3. Travel Considerations

a. Pick the easiest and safest route (non-combat).

b. Maintain a realistic pace; take rest stops when needed.

c. Avoid overdressing and overheating.

d. Consider food and water requirements.

e. Take special care of feet (change socks regularly).

f. Pack equipment to prevent loss, damage, pack imbalance, and personal safety.

g. Go **around** obstacles, not over or through them.

h. Travel on trails whenever possible (non-combat).

i. Travel in forested areas if possible.

j. Avoid creek bottoms and ravines with ***NO*** escape in the event of heavy rains.

k. Consider the following for swamps, lakes, and unfordable rivers:

(1) Circumnavigate swamps, lakes, and bogs if needed.

(2) Travel downstream to find people and slower water.

(3) Travel upstream to find narrower and shallow water.

4. River Travel

River travel may be faster and save energy when hypothermia is not a factor. It may be a primary mode of travel and LOC in a tropical environment **(use with caution if evading)**.

a. Use flotation device (raft, log, bamboo, etc.).

b. Use a pole to move the raft in shallow water.

c. Use an oar in deep water.

d. Stay near inside edge of river bends (current speed is less).

e. Keep near shore.

f. Watch for the following ***DANGERS:***

(1) Snags.

(2) Sweepers (overhanging limbs and trees).

(3) Rapids (***DO NOT*** attempt to shoot the rapids).

(4) Waterfalls.

(5) Hazardous animals.

g. Consider using a flotation device when crossing rivers or large/deep streams.

5. Ice and Snow Travel

Travel should be limited to areas free of hazards.

a. ***DO NOT*** travel in—

(1) Blizzards.

(2) Bitterly cold winds.

(3) Poor visibility.

b. Obstacles to winter travel follow:

(1) Reduced daylight hours (***BE AWARE***).

(2) Deep soft snow (if movement is necessary, make snowshoes **[Figure II-10])**. Travel is easier in early morning or late afternoon near dusk when snow is frozen or crusted.

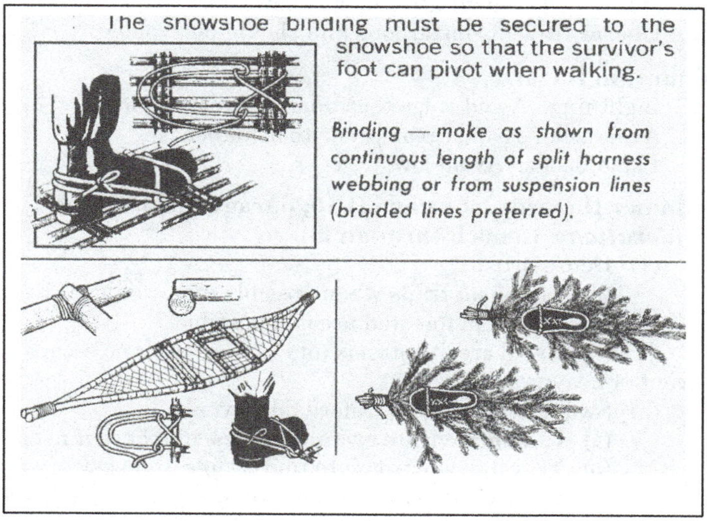

Figure II-10. Improvised Snowshoes

(3) Avalanche prone areas to avoid:

(a) Slopes 30-45 degrees or greater.

(b) Trees without uphill branches (identifies prior avalanches).

(c) Heavy snow loading on ridge tops.

(4) If caught in an avalanche, do the following:

(a) Backstroke to decrease burial depth.

(b) Move hand around face to create air pocket as moving snow slows.

(5) Frozen water crossings.

(a) Weak ice should be expected where—

- Rivers are straight.
- Objects protrude through ice.
- Snow banks extend over the ice.
- Rivers or streams come together.
- Water vapor rising indicates open or warm areas.

(b) Air pockets form when a frozen river loses volume.

(c) When crossing frozen water, distribute your weight by laying flat, belly crawling, or using snowshoes.

c. Glacier travel is hazardous and should be avoided.

6. Mountain Hazards

a. Lightning. Avoid ridge tops during thunderstorms.

b. Avalanche. Avoid areas prone to avalanches.

c. Flash floods. Avoid low areas.

7. Summer Hazards (see page II-10; paragraph 3, *Travel Considerations*, items h through k.)

(1) Dense brush.

(a) Travel on trails when possible (non-combat).

(b) Travel in forested areas if possible.

(c) Avoid creek bottoms and ravines with no escape in the event of heavy rains.

(2) Swamps, lakes, and unfordable rivers.

(a) Circumnavigate swamps, lakes, and bogs if needed.

(b) Travel downstream to find people and slower water.

(c) Travel upstream to find narrower and shallow water.

8. Dry Climates

a. ***DO NOT*** travel unless certain of reaching the destination using the water supply available.

b. Travel at dawn or dusk on hot days.

c. Follow the easiest trail possible (non-combat), avoiding—

(1) Deep sandy dune areas.

(2) Rough terrain.

d. In sand dune areas—

(1) Follow hard valley floor between dunes.

(2) Travel on the windward side of dune ridges.

e. If a sandstorm occurs—

(1) Mark your direction of travel.

(2) Sit or lie down in direction of travel.

(3) Try to get to the downwind side of natural shelter.

(4) Cover the mouth and nose with a piece of cloth.

(5) Protect the eyes.

(6) Remain stationary until the storm is over.

9. Tropical Climates

a. Travel only when it is light.

b. Avoid obstacles like thickets and swamps.

c. Part the vegetation to pass through. Avoid grabbing vegetation; it may have spines or thorns (**use gloves** if possible).

d. ***DO NOT*** climb over logs if you can go around them.

e. Find trails—

(1) Where **2** streams meet.

(2) Where a low pass goes over a range of hills.

f. While traveling trails—

(1) Watch for disturbed areas on game trails; they may indicate a pitfall or trap.

(2) Use a walking stick to probe for pitfalls or traps.

(3) ***DO NOT*** sleep on the trail.

(4) Exercise caution, the enemy uses the trails also.

10. Open Seas

a. Using currents—

(1) Deploy sea anchor **(Figure II-11)**. Sea anchor may be adjusted to make use of existing currents.

(2) Sit low in the raft.

(3) Deflate the raft slightly so it rides lower in the water.

b. Using winds—

(1) Pull in sea anchor.

(2) Inflate raft so it rides higher.

(3) Sit up in raft so body catches the wind.

(4) Construct a shade cover/sail **(Figure II-12)**. (Sail aids in making landfall.)

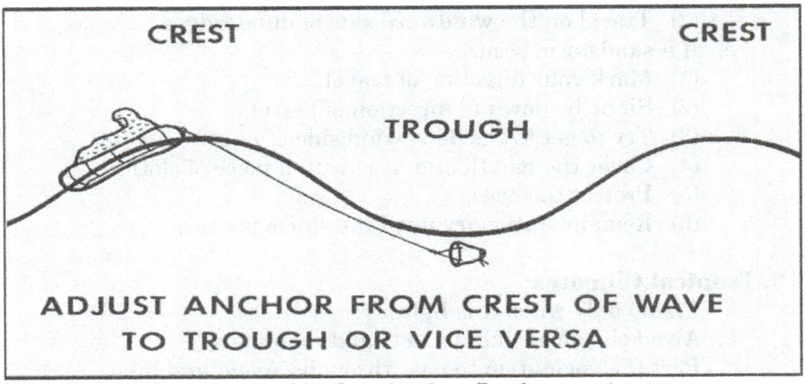

Figure II-11. Sea Anchor Deployment

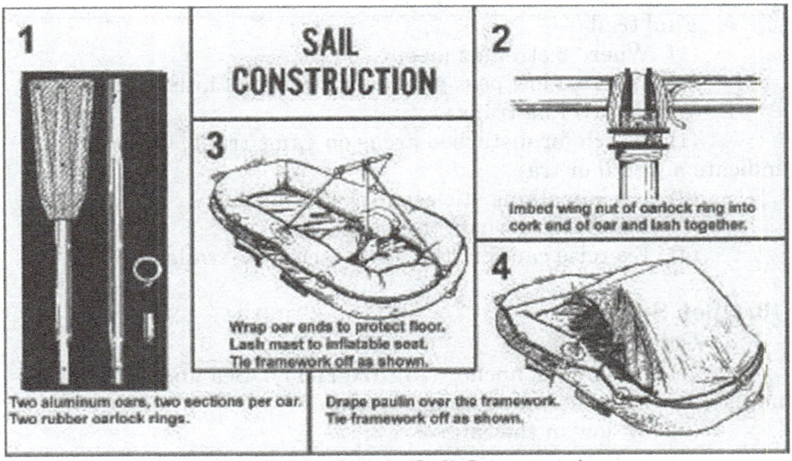

Figure II-12. Shade/Sail Construction

c. Making landfall. Indications of land are—

(1) Fixed cumulus clouds in a clear sky or in a cloudy sky where all other clouds are moving.

(2) Greenish tint in the sky **(in the tropics)**.

(3) Lighter colored reflection on clouds (open water causes dark gray reflections) **(in the arctic)**.

(4) Lighter colored water (indicates shallow water).

(5) The odors and sounds.

(a) Odors from swamps and smoke.

(b) Roar of surf/bird cries coming from one direction.

(6) Directional flights of birds at dawn and at dusk.

d. Swimming ashore—

(1) Consider physical condition.

(2) Use a flotation aid.

(3) Secure all gear to body before reaching landfall.

(4) Remain in raft as long as possible.

(5) Use the sidestroke or breaststroke to conserve strength if thrown from raft.

(6) Wear footgear and at least **1** layer of clothing.

(7) Try to make landfall during the lull between the sets of waves (waves are generally in **sets** of **7**, from **smallest** to **largest**).

(8) In moderate surf.

(a) Swim forward on the back of a wave.

(b) Make a shallow dive just before the wave breaks to end the ride.

(9) In high surf.

(a) Swim shoreward in the trough between waves.

(b) When the seaward wave approaches, face it and submerge.

(c) After it passes, work shoreward in the next trough.

(10) If caught in the undertow of a large wave—

(a) Remain calm and swim to the surface.

(b) Lie as close to the surface as possible.

(c) Parallel shoreline and attempt landfall at a point further down shore.

(11) Select a landing point.

(a) Avoid places where waves explode upon rocks.

(b) Find a place where waves smoothly rush onto the rocks.

(12) After selecting a landing site—

(a) Face shoreward.

(b) Assume a sitting position with feet 2 or 3 feet lower than head to absorb the shock of hitting submerged objects.

e. Rafting ashore—

(1) Select landing point carefully.

(2) Use caution landing when the sun is low and straight in front of you causing poor visibility.

(3) Land on the lee (downwind) side of islands or point of land if possible.

(4) Head for gaps in the surf line.

(5) Penetrate surf by—

(a) Taking down most shade/sails.

(b) Using paddles to maintain control.

(c) Deploying a sea anchor for stability.

CAUTION: ***DO NOT*** deploy a sea anchor if traveling through coral.

f. Making sea ice landings on large stable ice flows. Icebergs, small flows, and disintegrating flows are dangerous **(ice can cut a raft)**.

(1) Use paddles to avoid sharp edges.

(2) Store raft away from the ice edge.

(3) Keep raft inflated and ready for use.

(4) Weight down/secure raft so it does not blow away.

Chapter III

RADIO COMMUNICATIONS AND SIGNALING

Inventory and review the operating instructions of all communications and signaling equipment.

1. Radio Communications (Voice and Data)

a. Non-combat.

(1) Ensure locator beacon is operational.

(2) Follow standing plans for on/off operations to conserve battery use.

b. Combat.

(1) Turn off locator beacon.

(2) Keep it with you to supplement radio communications.

(3) Follow plans/orders for on/off operations.

c. Make initial contact as soon as possible or as directed in applicable plans/orders.

d. If no immediate contact, then as directed in applicable plans/orders.

e. Locate spare radio and batteries (keep warm and dry).

f. Transmissions.

(1) Use concealment sites (combat) that optimize line of site (LOS).

(2) Face recovery asset.

(3) Keep antenna perpendicular to intended receiver **(Figure III-1)**.

(4) ***DO NOT*** ground antenna (that is finger on antenna or attaching bolt, space blanket, vegetation, etc.).

(5) Keep transmissions short (3-5 seconds maximum). Use data burst if available.

(6) Move after each transmission (***ONLY*** in combat, if possible).

(7) If transmitting in the blind, ensure a clear LOS towards the equator.

(8) Use terrain masking to hinder enemy direction finding.

g. Listening (use reception times in applicable plans/orders or as directed by recovery forces).

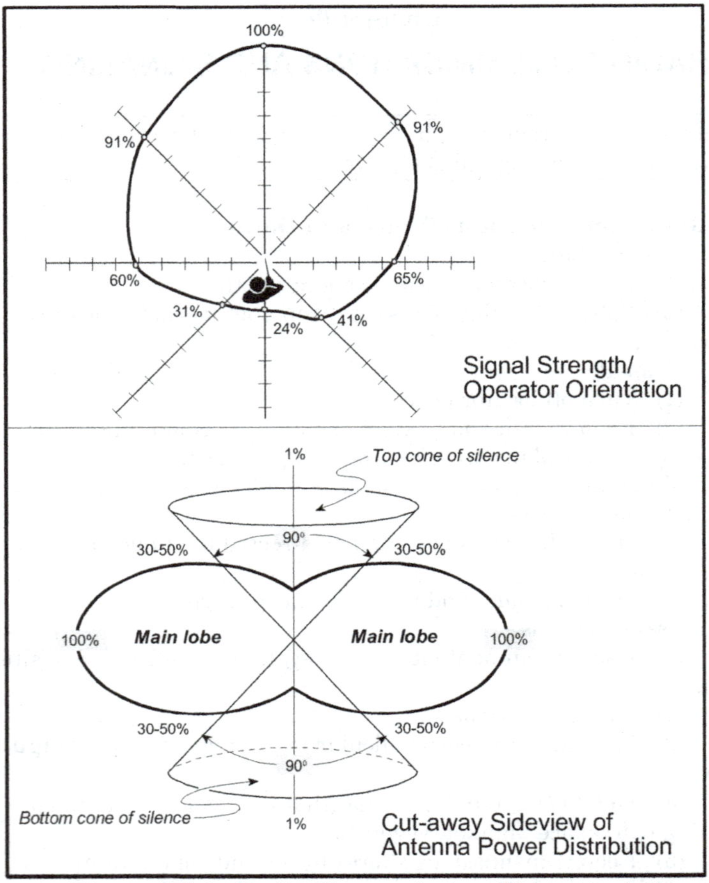

Figure III-1. Radio Transmission Characteristics

2. Signaling

a. Pyrotechnic signals.

(1) Prepare early (weather permitting).

(2) Use as directed in applicable plans/orders or as directed by recovery forces.

(3) Extend over raft's edge before activating.

b. Signal mirror **(Figure III-2)**.
   (1) Use as directed by recovery forces.
   (2) If no radio, use only with confirmed friendly forces.
   (3) Cover when not in use.

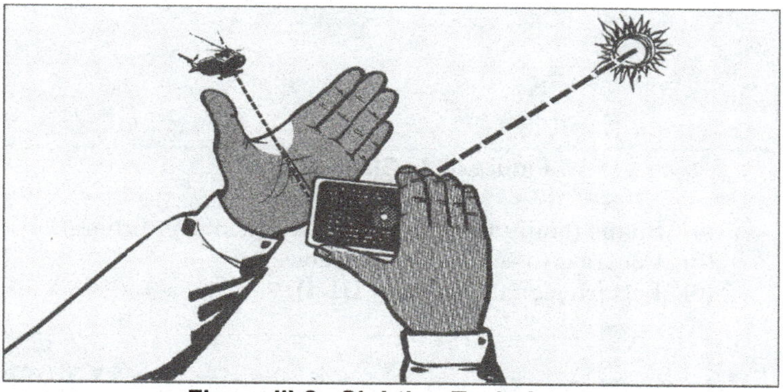

Figure III-2. Sighting Techniques

> **Note:** Make a mirror from any shiny metal or glass.

c. Strobe/IR lights.
   (1) Prepare early, consider filters and shields.
   (2) Use as directed by recovery forces.
   (3) Conserve battery life.

> **Note:** Produces one residual flash when turned off.

d. Pattern signals (use as directed in applicable plans/orders).
   (1) Materials:
      (a) Manmade (space blanket, signal paulin, parachute).
      (b) Natural use materials that contrast the color and/or texture of the signaling area (rocks, brush, branches, stomped grass).
   (2) Location.
      (a) Maximize visibility from above.
      (b) Provide concealment from ground observation.
   (3) Size (large as possible) and ratio **(Figure III-3)**.

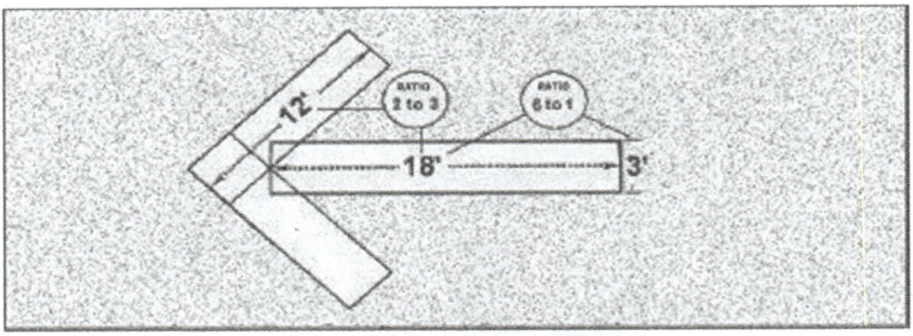

Figure III-3. Size and Ratio

(4) Shape (maintain straight lines and sharp corners).
(5) Contrast (use color and shadows).
(6) Pattern signals **(Figure III-4)**.

NO.	MESSAGE	CODE SYMBOL
1	REQUIRE ASSISTANCE	V
2	REQUIRE MEDICAL ASSISTANCE	X
3	NO or NEGATIVE	N
4	YES or AFFIRMATIVE	Y
5	PROCEEDING IN THIS DIRECTION	↑

Figure III-4. Signal Key

e. Sea dye marker.
(1) ***DO NOT*** waste in rough seas or fast moving water.
(2) Conserve unused dye by rewrapping.
(3) May be used to color snow.

f. Non-combat considerations:

(1) Use a fire at night.

(2) Use smoke for day (tires or petroleum products for dark smoke and green vegetation for light smoke). **(Figure III-5)**

(3) Use signal mirror to sweep horizon.

(4) Use audio signals (that is, voice, whistle, and weapons fire).

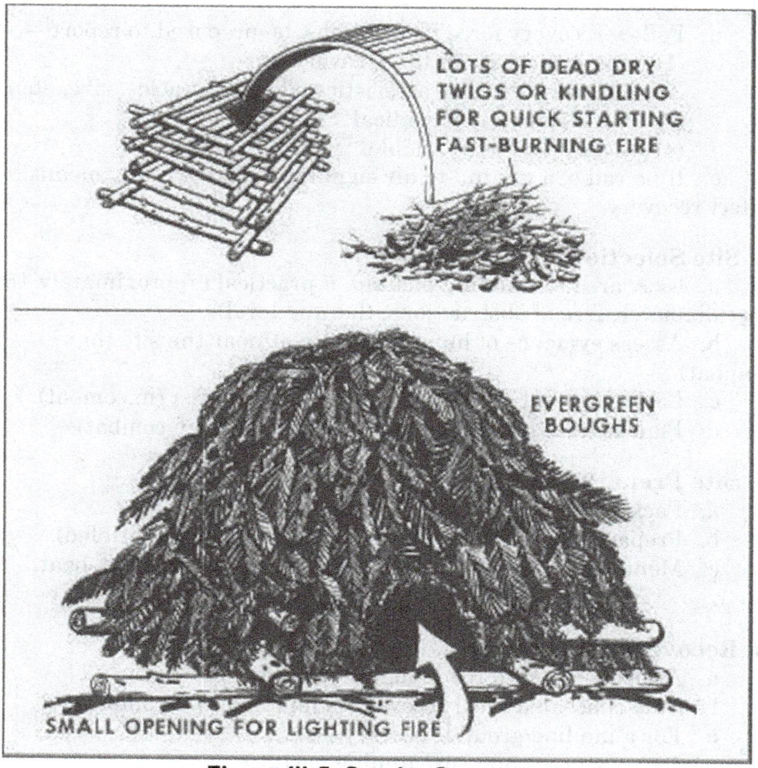

Figure III-5. Smoke Generator

Chapter IV
RECOVERY

1. Responsibilities

a. Establish radio contact with recovery forces (if possible).

b. Maintain communication with recovery forces until recovered.

c. Be prepared to authenticate as directed in applicable plans/orders.

d. Follow recovery force instructions, be prepared to report—

(1) Enemy activity in the recovery area.

(2) Recovery site characteristics (slope, obstacles, size, etc.).

(3) Number in party/medical situation.

(4) Signal devices available.

e. If no radio, a ground-to-air signal may be your only means to effect recovery.

2. Site Selection

a. Locate area for landing pick-up, if practical (approximately 150 feet diameter, free of obstructions, flat and level).

b. Assess evidence of human activity at/near the site (in combat).

c. Locate several concealment sites around area (in combat).

d. Plan several tactical entry and exit routes (in combat).

3. Site Preparation

a. Pack and secure all equipment.

b. Prepare signaling devices (use as directed or as briefed).

c. Mentally review recovery methods (aircraft, ground, boat, etc.).

4. Recovery Procedures

a. Assist recovery force in identifying your position.

b. Stay concealed until recovery is imminent (in combat).

c. For a landing/ground recovery—

(1) Assume a non-threatening posture.

(2) Secure weapons and avoid quick movement.

(3) ***DO NOT*** approach recovery vehicle until instructed.

(4) Beware of rotors/propellers when approaching recovery vehicle, especially on sloping or uneven terrain. Secure loose equipment that could be caught in rotors/propellers.

d. For **hoist** recovery devices **(Figures IV-1 and IV-2)**—

(1) Use eye protection, if available (glasses or helmet visor).

(2) Allow metal on device to contact the surface before touching to avoid injury from static discharge.

(3) Sit or kneel for stability while donning device.

(4) Put safety strap under armpits.

(5) Ensure cable is in front of you.

(6) Keep hands clear of all hardware and connectors.

(7) ***DO NOT*** become entangled in cable.

(8) Use a thumbs up, vigorous cable shake, or radio call to signal you are ready.

(9) Drag feet on the ground to decrease oscillation.

(10) ***DO NOT*** assist during hoist or when pulled into the rescue vehicle. Follow crewmember instructions.

e. For **nonhoist** recovery (rope or unfamiliar equipment)—

(1) Create a ***"fixed loop"*** big enough to place under armpits **(Figure IV-3)**.

(2) Follow the procedures in "**d**" above.

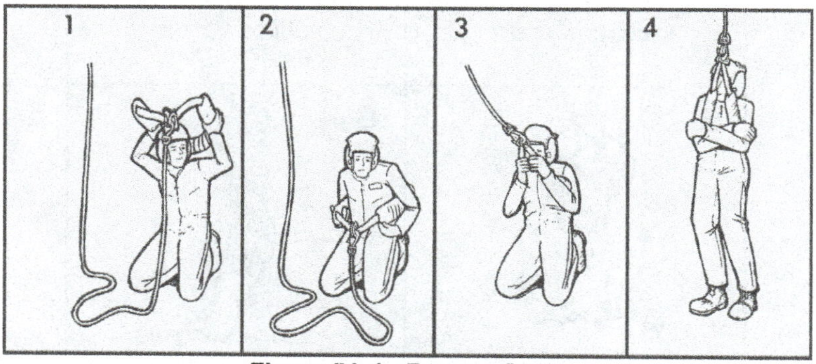

Figure IV–1. Rescue Strap

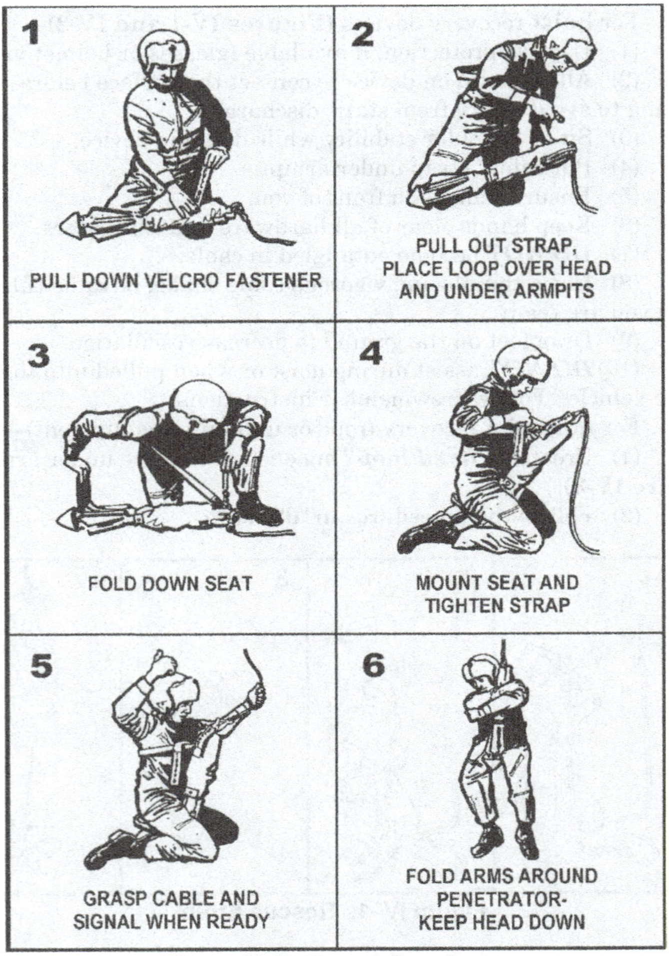

Figure IV-2. Forest Penetrator

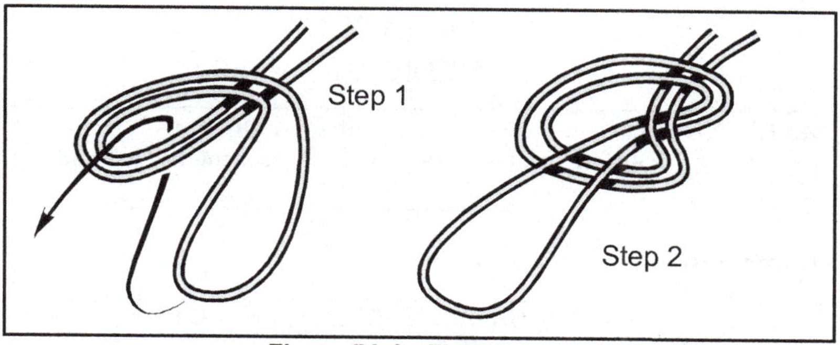

Figure IV–3. Fixed Loop

Chapter V
MEDICAL

WARNING: These emergency medical procedures are for survival situations. Obtain professional medical treatment as soon as possible.

1. Immediate First Aid Actions

Remember the *ABCs* of Emergency Care:
Airway Breathing Circulation

a. Determine responsiveness as follows:
   (1) If unconscious, arouse by shaking gently and shouting.
   (2) If no response—
      (a) Keep head and neck aligned with body.
      (b) Roll victims onto their backs.
      (c) Open the airway by lifting the chin **(Figure V-1)**.
      (d) Look, listen, and feel for air exchange.

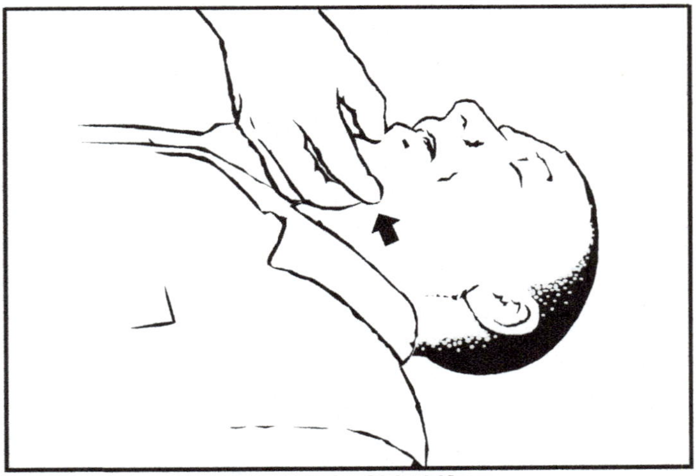

Figure V-1. Chin Lift

(3) If victim is not breathing—

(a) Check for a clear airway; remove any blockage.

(b) Cover victim's mouth with your own.

(c) Pinch victim's nostrils closed.

(d) Fill victim's lungs with **2** slow breaths.

(e) If breaths are blocked, reposition airway; try again.

(f) If breaths still blocked, give **5** abdominal thrusts:

- Straddle the victim.
- Place a fist between breastbone and belly button.
- Thrust upward to expel air from stomach.

(g) Sweep with finger to clear mouth.

(h) Try **2** slow breaths again.

(i) If the airway is still blocked, continue **(c)** through **(f)** until successful or exhausted.

(j) With open airway, start mouth to mouth breathing:

- Give **1** breath every 5 seconds.
- Check for chest rise each time.

(4) If victim is unconscious, but breathing—

(a) Keep head and neck aligned with body.

(b) Roll victim on side (drains the mouth and prevents the tongue from blocking airway).

(5) If breathing difficulty is caused by chest trauma, refer to **page V-7, paragraph 1d, *Treat Chest Injuries***.

CAUTION: ***DO NOT*** remove an impaled object unless it interferes with the airway. You may cause more tissue damage and increase bleeding. For travel, you may shorten and secure the object.

b. Control bleeding as follows:

(1) Apply a pressure dressing **(Figure V-2)**.

(2) If ***STILL*** bleeding—

(a) Use direct pressure over the wound.

(b) Elevate the wounded area above the heart.

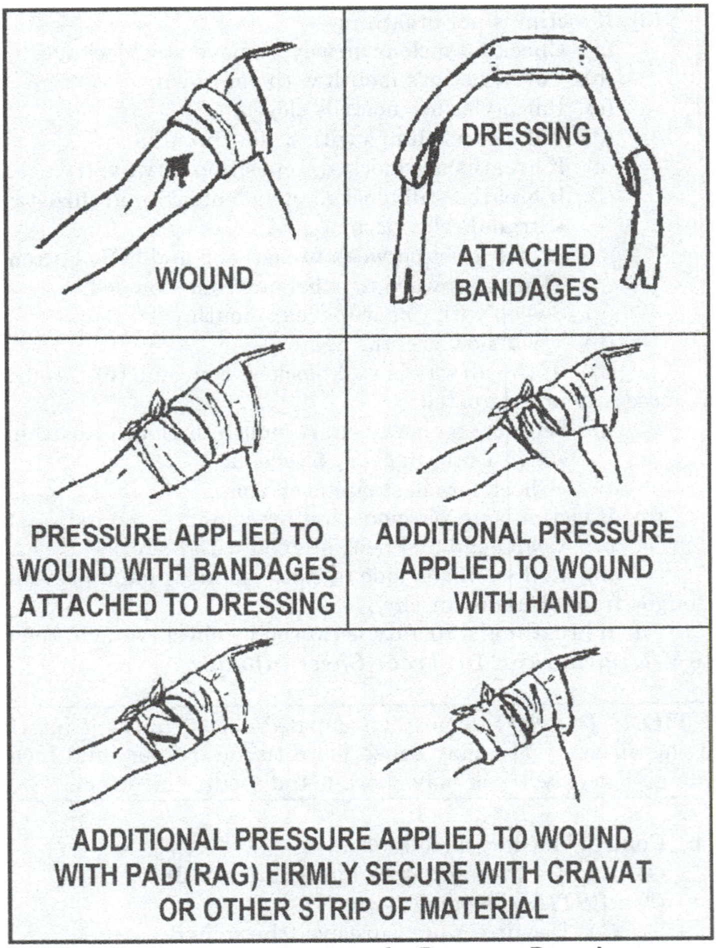

Figure V-2. Application of a Pressure Dressing

(3) If ***STILL*** bleeding—

(a) Use a pressure point between the injury and the heart **(Figure V-3)**.

(b) Maintain pressure for 6 to 10 minutes before checking to see if bleeding has stopped.

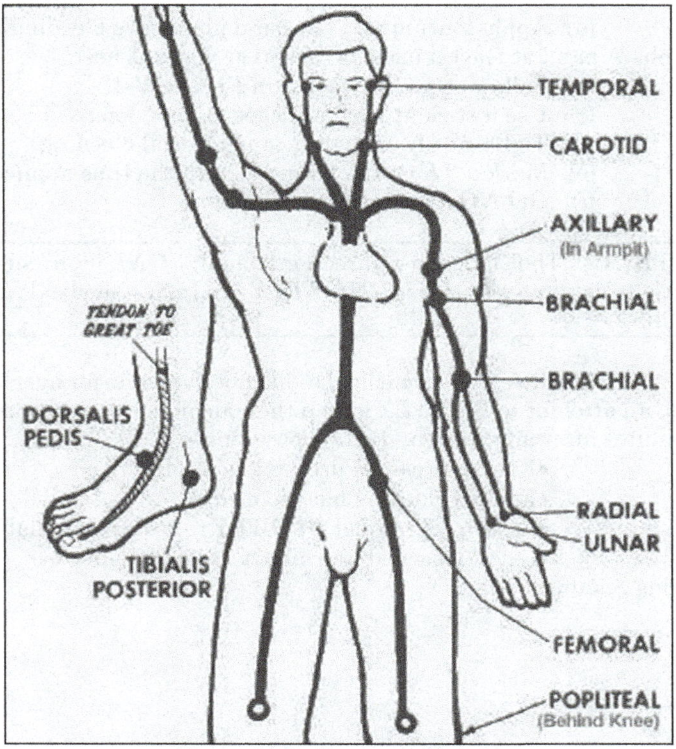

Figure V-3. Pressure Points

(4) If a limb wound is ***STILL*** bleeding—

CAUTION: Use of a tourniquet is a ***LAST RESORT*** measure. Use ***ONLY*** when severe, uncontrolled bleeding will cause loss of life. Recognize that long-term use of a tourniquet may cause loss of limb.

(a) Apply tourniquet (TK) band just above bleeding site on limb. A band at least 3 inches (7.5 cm) or wider is best.

(b) Follow steps illustrated in **Figure V-4**.

(c) Use a stick at least 6 inches (15 cm) long.

(d) Tighten only enough to stop arterial bleeding.

(e) Mark a ***TK*** on the forehead with the time applied.

(f) ***DO NOT*** cover the tourniquet.

CAUTION: The following directions apply ***ONLY*** in survival situations where rescue is ***UNLIKELY*** and ***NO*** medical aid is available.

(g) If rescue or medical aid is not available for over 2 hours, an attempt to ***SLOWLY*** loosen the tourniquet may be made 20 minutes after application. Before loosening—

- Ensure pressure dressing is in place.
- Ensure bleeding has stopped
- Loosen tourniquet ***SLOWLY*** to restore circulation.
- Leave loosened tourniquet in position in case bleeding resumes.

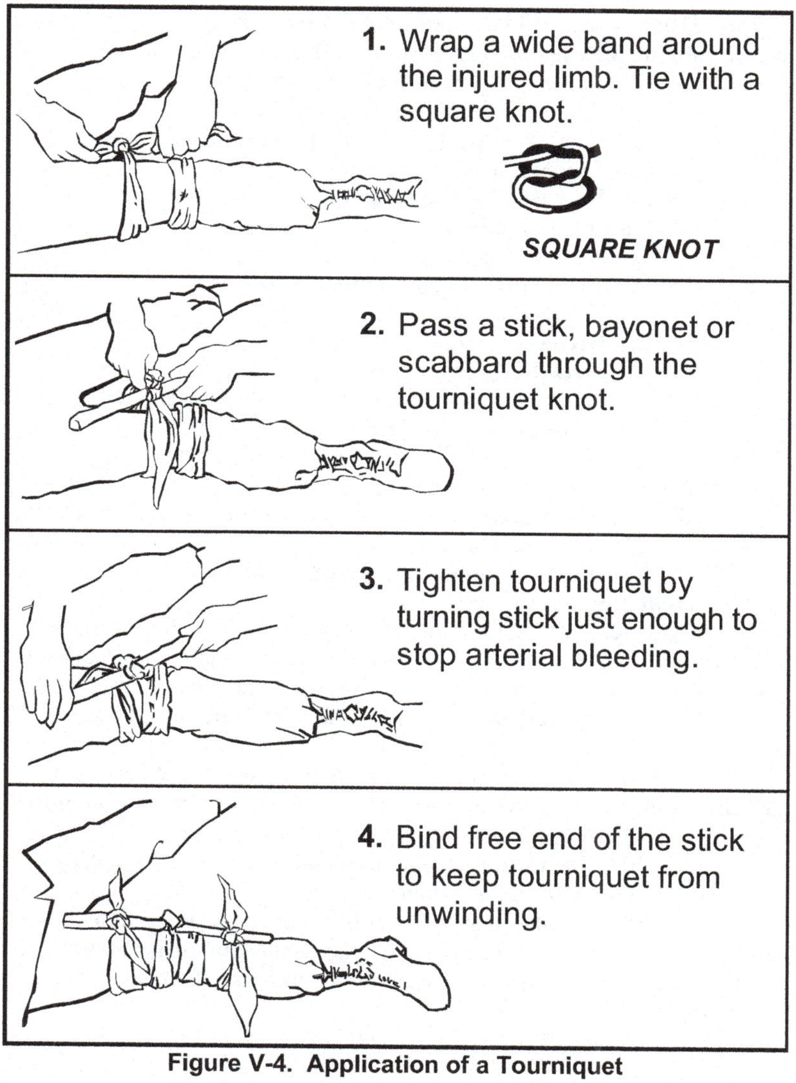

Figure V-4. Application of a Tourniquet

c. Treat shock. (Shock is difficult to identify or treat under field conditions. It may be present with or without visible injury.)

(1) Identify by one or more of the following:

(a) Pale, cool, and sweaty skin.

(b) Fast breathing and a weak, fast pulse.

(c) Anxiety or mental confusion.

(d) Decreased urine output.

(2) Maintain circulation.

(3) Treat underlying injury.

(4) Maintain normal body temperature.

(a) Remove wet clothing.

(b) Give warm fluids.

•***DO NOT*** give fluids to an unconscious victim.

•***DO NOT*** give fluids if they cause victim to gag.

(c) Insulate from ground.

(d) Shelter from the elements.

(5) Place conscious victim on back.

(6) Place very weak or unconscious victim on side, this will—

(a) Allow mouth to drain.

(b) Prevent tongue from blocking airway.

d. Treat chest injuries.

(1) Sucking chest wound. This occurs when chest wall is penetrated; may cause victim to gasp for breath; may cause sucking sound; may create bloody froth as air escapes the chest.

(a) ***Immediately*** seal wound with hand or airtight material.

(b) Tape airtight material over wound on ***3 sides only*** **(Figure V-5)** to allow air to escape from the wound but not to enter.

(c) Monitor breathing and check dressing.

(d) Lift untapped side of dressing as victim ***exhales*** to allow trapped air to escape, as necessary.

(2) Flail chest. Results from blunt trauma when ***3*** or ***more*** ribs are broken in ***2*** or more places. The flail segment is the broken area that moves in a direction opposite to the rest of chest during breathing.

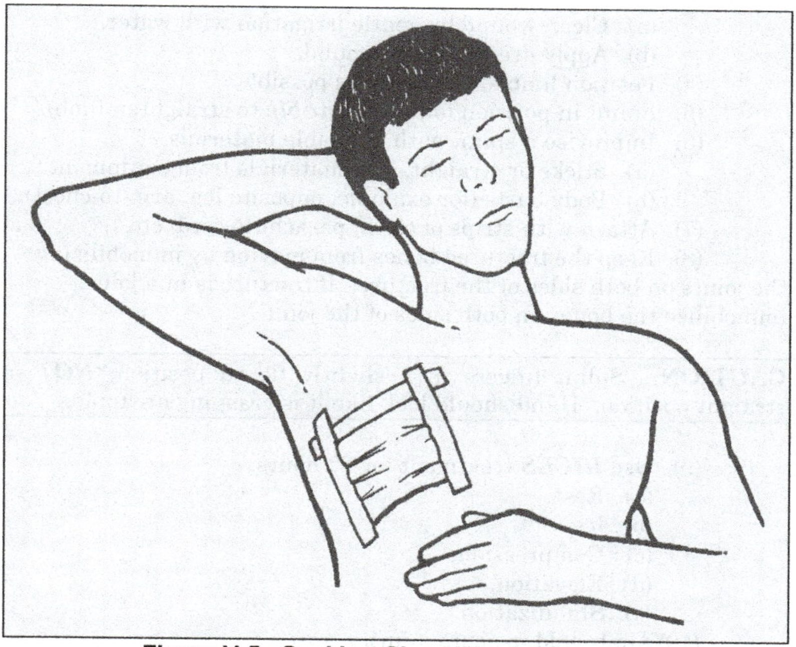

Figure V-5. Sucking Chest Wound Dressing

(a) Stabilize the flail segment as follows:
- Place rolled-up clothing or bulky pad over site.
- Tape pad to site
- ***DO NOT*** wrap tape around chest.

(b) Have victim keep segment still with hand pressure.

(c) Roll victim onto side of flail segment injury (as other injuries allow).

(3) Fractured ribs.

(a) Encourage deep breathing (painful, but necessary to prevent the possible development of pneumonia).

(b) ***DO NOT*** constrict breathing by taping ribs.

e. Treat fractures, sprains, and dislocations.

(1) Control bleeding.

(2) Remove watches, jewelry, and constrictive clothing.

(3) If fracture penetrates the skin—

(a) Clean wound by gentle irrigation with water.
(b) Apply dressing over wound.
(4) Position limb as normally as possible.
(5) Splint in position found (if ***unable*** to straighten limb).
(6) Improvise a splint with available materials:
(a) Sticks or straight, stiff materials from equipment.
(b) Body parts (for example, opposite leg, arm-to-chest).
(7) Attach with strips of cloth, parachute cord, etc.
(8) Keep the fractured bones from moving by immobilizing the joints on both sides of the fracture. If fracture is in a joint, immobilize the bones on both sides of the joint.

CAUTION: Splint fingers in a slightly flexed position, ***NOT*** in straight position. Hand should look like it is grasping an apple.

(9) Use ***RICES*** treatment for 72 hours.
(a) **R**est.
(b) **I**ce.
(c) **C**ompression.
(d) **E**levation.
(e) **S**tabilization.
(10) Apply cold to acute injuries.
(11) Use 15 to 20 minute periods of cold application.
(a) ***DO NOT*** use continuous cold therapy.
(b) Repeat 3 to 4 times per day.
(c) Avoid cooling that can cause frostbite or hypothermia.
(12) Wrap with a compression bandage after cold therapy.
(13) Elevate injured area above heart level to reduce swelling.
(14) Check periodically for a pulse beyond the injury site.
(15) Loosen bandage or reapply splint if no pulse is felt or if swelling occurs because bandage is too tight.

2. Common Injuries and Illnesses

a. Burns.
(1) Cool the burned area with water.
(a) Use immersion or cool compresses.
(b) Avoid aggressive cooling with ice or frigid water.
(2) Remove watches, jewelry, constrictive clothing.

(3) ***DO NOT*** remove embedded, charred material that will cause burned areas to bleed.

(4) Cover with sterile dressings.

(5) ***DO NOT*** use lotion or grease.

(6) Avoid moving or rubbing the burned part.

(7) Drink ***extra*** water to compensate for increased fluid loss from burns. (Add ***1/4 teaspoon*** of ***salt*** [if available] to ***each quart*** of ***water***.)

(8) Change dressings when soaked or dirty.

b. Eye injuries.

(1) Sun/snow blindness (gritty, burning sensation, and possible reduction in vision caused by sun exposure).

(a) Prevent with improvised goggles. **(See Chapter VI, page VI-3, Figure VI-2.)**

(b) Treat by patching affected eye(s).

- Check after 12 hours.
- Replace patch for another 12 hours if not healed.

(c) Use cool compresses to reduce pain.

(2) Foreign body in eye.

(a) Irrigate with clean water from the ***inside*** to the ***outside*** corner of the eye.

(b) If foreign body is not removed by irrigation, improvise a small swab. Moisten and wipe gently over the affected area.

(c) If foreign body is ***STILL*** not removed, patch eye for 24 hours and then reattempt removal using steps **(a)** and **(b)**.

c. Heat injury.

(1) Heat cramps (cramps in legs or abdomen).

(a) Rest.

(b) Drink water. Add ***1/4 teaspoon*** of salt ***per quart***.

(2) Heat exhaustion (pale, sweating, moist, cool skin).

(a) Rest in shade.

(b) Drink water.

(c) Protect from further heat exposure.

(3) Heat stroke (victim disoriented or unconscious, skin is hot and flushed [sweating **may** or **may not** occur], fast pulse).

CAUTION: Handle heat stroke victim gently. Shock, seizures, and cardiac arrest can occur.

(a) Cool as rapidly as possible (saturate clothing with water and fan the victim). Remember to cool the groin and armpit areas. (Avoid overcooling.)

(b) Maintain airway, breathing, and circulation.

d. Cold injuries:

(1) Frostnip and frostbite—

(a) Are progressive injuries.

•Ears, nose, fingers, and toes are affected first.

•Areas will feel cold and may tingle leading to—

••Numbness that progresses to—

•••Waxy appearance with stiff skin that cannot glide freely over a joint.

(b) Frostnipped areas rewarm with body heat. If body heat ***WILL NOT*** rewarm area in 15 to 20 minutes, then frostbite is present.

(c) Frostbitten areas are deeply frozen and require medical treatment.

CAUTION: In frostbite, repeated freezing and thawing causes severe pain and increases damage to the tissue. ***DO NOT*** rub frozen tissue. ***DO NOT*** thaw frozen tissue.

(2) Hypothermia—

(a) Is a progressive injury.

•Intense shivering with impaired ability to perform complex tasks leads to—

••Violent shivering, difficulty speaking, sluggish thinking go to—

•••Muscular rigidity with blue, puffy skin; jerky movements go to—

••••Coma, respiratory and cardiac failure.

(b) Protect victim from the environment as follows:

•Remove wet clothing.

•Put on dry clothing (if available).

•Prevent further heat loss.

••Cover top of head.

••Insulate from above and below.

•Warm with blankets, sleeping bags, or shelter.

•Warm central areas before extremities.

••Place heat packs in groin, armpits, and around neck.

••Avoid causing burns to skin.

CAUTION: Handle hypothermia victim gently. Avoid overly rapid rewarming which may cause cardiac arrest. Rewarming of victim with skin-to-skin contact by volunteer(s) inside of a sleeping bag is a survival technique but can cause internal temperatures of all to drop.

e. Skin tissue damage.

(1) Immersion injuries. Skin becomes wrinkled as in ***dishpan hands***.

(a) Avoid walking on affected feet.
(b) Pat dry; ***DO NOT*** rub. Skin tissue will be sensitive.
(c) Dry socks and shoes. Keep feet protected.
(d) Loosen boots, cuffs, etc., to improve circulation.
(e) Keep area dry, warm, and open to air.
(f) ***DO NOT*** apply creams or ointments.

(2) Saltwater sores.

(a) Change body positions frequently.
(b) Keep sores dry.
(c) Use antiseptic (if available).
(d) ***DO NOT*** open or squeeze sores.

f. Snakebite.

CAUTION: This snakebite treatment recommendation is for situations where medical aid and specialized equipment are not available.

(1) Nonpoisonous. Clean and bandage wound.

(2) Poisonous.

(a) Remove constricting items.
(b) Minimize activity.
(c) ***DO NOT*** cut the bite site; ***DO NOT*** use your mouth to create suction.
(d) Clean bite with soap and water; cover with a dressing.

(e) Overwrap the bite site with a tight (elastic) bandage **(Figure V-6)**. The intent is to slow capillary and venous blood flow but not arterial flow. Check for pulse below the overwrap.

(f) Splint bitten extremity to prevent motion.

(g) Treat for shock **(page V-7, paragraph 1c)**.

(h) Position extremity below level of heart.

(i) Construct shelter if necessary (let the victim rest).

(j) For conscious victims, force fluids.

g. Marine life.

(1) Stings.

(a) Flush wound with salt water (fresh water stimulates toxin release).

(b) Remove jewelry and watches.

(c) Remove tentacles and gently scrape or shave skin.

(d) Apply a steroid cream (if available).

(e) ***DO NOT*** rub area with sand.

(f) Treat for shock; artificial respiration may be required **(page V-1, paragraph 1a)**.

(g) ***DO NOT*** use urine to flush or treat wounds.

(2) Punctures.

(a) Immerse affected part in hot water or apply hot compresses for 30-60 minutes (as hot as victim can tolerate).

(b) Cover with clean dressing.

(c) Treat for shock as needed.

h. Skin irritants (includes poison oak and poison ivy).

(1) Wash with large amounts of water. Use soap (if available).

(2) Keep covered to prevent scratching.

i. Infection.

(1) Keep wound clean.

(2) Use iodine tablet solution or diluted betadine to prevent or treat infection.

(3) Change bandages as needed.

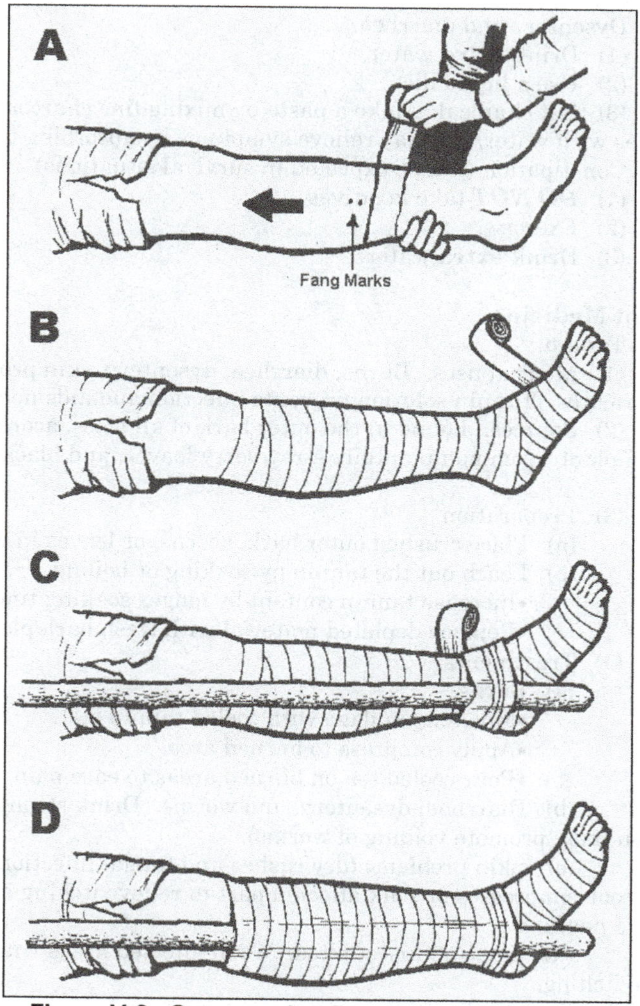

Figure V-6. Compression Bandage for Snake Bite

j. Dysentery and diarrhea.

(1) Drink ***extra*** water.

(2) Use a liquid diet.

(3) Eat charcoal. Make a paste by mixing fine charcoal particles with water. (It may relieve symptoms by absorbing toxins.)

k. Constipation (can be expected in survival situations).

(1) ***DO NOT*** take laxatives.

(2) Exercise.

(3) Drink **extra** water.

3. Plant Medicine

a. Tannin.

(1) Medical uses. Burns, diarrhea, dysentery, skin problems, and parasites. Tannin solution prevents infection and aids healing.

(2) Sources. Found in the outer bark of all trees, acorns, banana plants, common plantain, strawberry leaves, and blackberry stems.

(3) Preparation.

(a) Place crushed outer bark, acorns, or leaves in water.

(b) Leach out the tannin by soaking or boiling.

•Increase tannin content by longer soaking time.

•Replace depleted material with fresh bark/plants.

(4) Treatments.

(a) Burns.

•Moisten bandage with cooled tannin tea.

•Apply compress to burned area.

•Pour cooled tea on burned areas to ease pain.

(b) Diarrhea, dysentery, and worms. Drink strong tea solution (may promote voiding of worms).

(c) Skin problems (dry rashes and fungal infections). Apply cool compresses or soak affected part to relieve itching and promote healing.

(d) Lice and insect bites. Wash affected areas with tea to ease itching.

b. Salicin/salicylic acid.

(1) Medical uses. Aches, colds, fever, inflammation, pain, sprains, and sore throat (aspirin-like qualities).

(2) Sources. Willow and aspen trees **(Figure V-7)**.

(3) Preparation.

(a) Gather twigs, buds, or cambium layer (soft, moist layer between the outer bark and the wood) of willow or aspen.

(b) Prepare tea as described in paragraph **3a(3)**.

(c) Make poultice.

•Crush the plant or stems.

•Make a pulpy mass.

(4) Treatments.

(a) Chew on twigs, buds, or cambium for symptom relief.

(b) Drink tea for colds and sore throat.

(c) Use warm, moist poultice for aches and sprains.

•Apply pulpy mass over injury.

•Hold in place with a dressing.

c. Common plantain.

(1) Medical uses. Itching, wounds, abrasions, stings, diarrhea, and dysentery.

(2) Source. There are over 200 plantain species with similar medicinal properties. The common plantain is shown in **Figure V-7**.

(3) Preparation.

(a) Brew tea from seeds.

(b) Brew tea from leaves.

(c) Make poultice of leaves.

(4) Treatments.

(a) Drink tea made from seeds for diarrhea or dysentery.

(b) Drink tea made from leaves for vitamin and minerals.

(c) Use poultice to treat cuts, sores, burns, and stings.

d. Papain.

(1) Medical uses. Digestive aid, meat tenderizer, and a food source.

(2) Source. Fruit of the papaya tree **(Figure V-7)**.

(3) Preparation.

(a) Make cuts in **unripe** fruit.

(b) Gather milky white sap for its papain content.

(c) Avoid getting sap in eyes or wounds.

(4) Treatments.

(a) Use sap to tenderize tough meat.

(b) Eat **ripe** fruit for food, vitamins, and minerals.

e. Common Cattail.

(1) Medical uses. Wounds, sores, boils, inflammations, burns, and an excellent food source.

(2) Source. Cattail plant found in marshes **(Figure V-7)**.

(3) Preparation.

(a) Pound roots into a pulpy mass for a poultice.

(b) Cook and eat green bloom spikes.

(c) Collect yellow pollen for flour substitute.

(d) Peel and eat tender shoots (raw or cooked).

(4) Treatments.

(a) Apply poultice to affected area.

(b) Use plant for food, vitamins, and minerals.

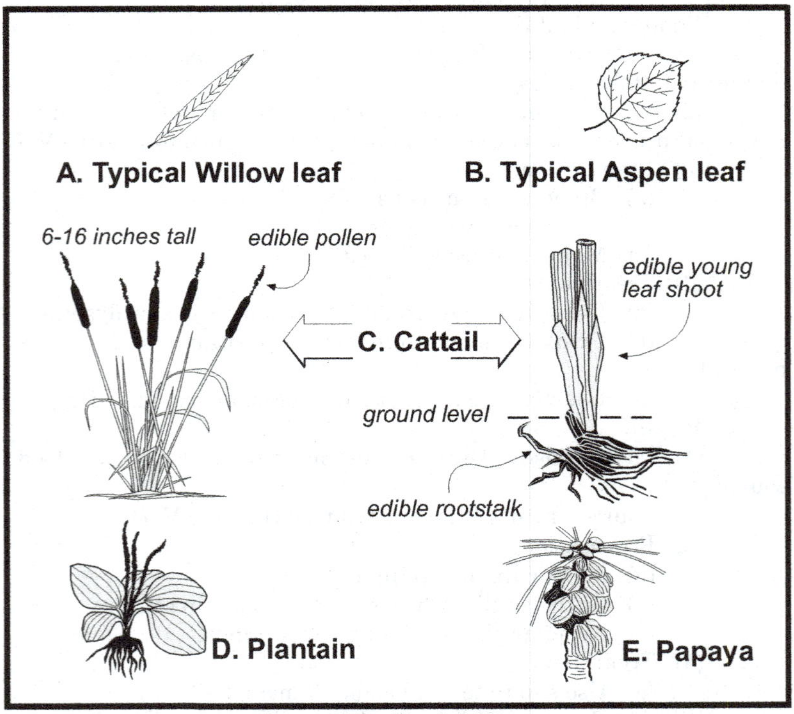

Figure V-7. Useful Plants

4. Health and Hygiene

a. Stay clean (daily regimen).

(1) Minimize infection by washing. (Use white ashes, sand, or loamy soil as soap substitutes.)

(2) Comb and clean debris from hair.

(3) Cleanse mouth and brush teeth.

(a) Use hardwood twig as toothbrush (fray it by chewing on one end then use as brush).

(b) Use single strand of an inner core string from parachute cord for dental floss.

(c) Use clean finger to stimulate gum tissues by rubbing.

(d) Gargle with salt water to help prevent sore throat and aid in cleaning teeth and gums.

(4) Clean and protect feet.

(a) Change and wash socks

(b) Wash, dry, and massage.

(c) Check frequently for blisters and red areas.

(d) Use adhesive tape/mole skin to prevent damage.

b. Exercise daily.

c. Prevent and control parasites.

(1) Check body for lice, fleas, ticks, etc.

(a) Check body regularly.

(b) Pick off insects and eggs (***DO NOT*** crush).

(2) Wash clothing and use repellents.

(3) Use smoke to fumigate clothing and equipment.

5. Rules for Avoiding Illness

a. Purify all water obtained from natural sources by using iodine tablets, bleach, or boiling for 5 minutes.

b. Locate latrines 200 feet from water and away from shelter.

c. Wash hands before preparing food or water.

d. Clean all eating utensils after each meal.

e. Prevent insect bites by using repellent, netting, and clothing.

f. Dry wet clothing as soon as possible.

g. Eat varied diet.

h. Try to get 7-8 hours sleep per day.

Chapter VI
PERSONAL PROTECTION

1. Priorities

a. Evaluate available resources and situation, then accomplish individual tasks accordingly.

b. First 24 hours in order of situational needs—

(1) Construct survival shelter according to selection criteria.

(2) Procure water.

(3) Establish multiple survival signals.

(4) Build Fire.

c. Second 24 hours in order of situational needs—

(1) Construct necessary tools and weapons.

(2) Procure food.

2. Care and Use of Clothing

a. Never discard clothing.

b. Wear loose and layered clothing.

(1) Tight clothing restricts blood flow regulating body temperature.

(2) Layers create more dead air space.

c. Keep entire body covered to prevent sunburn and dehydration in hot climates. When fully clothed, the majority of body heat escapes through the head and neck areas.

d. Avoid overheating.

(1) Remove layers of clothing before strenuous activities.

(2) Use a hat to regulate body heat.

(3) Wear a hat when in direct sunlight (in hot environment).

e. Dampen clothing when on the ocean in hot weather.

(1) Use salt water, ***NOT*** drinking water.

(2) Dry clothing before dark to prevent hypothermia.

f. Keep clothing dry to maintain its insulation qualities (dry damp clothing in the sun or by a fire).

g. If you fall into the water in the winter—

(1) Build fire.

(2) Remove wet clothing and rewarm by fire.

(3) Finish drying clothing by fire.

h. If no fire is available—

(1) Remove clothing and get into sleeping bag (if available).

(2) Allow wet clothes to freeze.

(3) Break ice out of clothing.

i. Keep clothing clean (dirt reduces its insulation qualities). Examine clothing frequently for damage.

(1) ***DO NOT*** sit or lie directly on the ground.

(2) Wash clothing whenever possible.

(3) Repair when necessary by using—

(a) Needle and thread.

(b) Safety pins.

(c) Tape.

j. Improvised foot protection **(Figure VI-1)**.

(1) Cut **2** to **4** layers of cloth into a 30-inch square.

(2) Fold into a triangle.

(3) Center foot on triangle with toes toward corner.

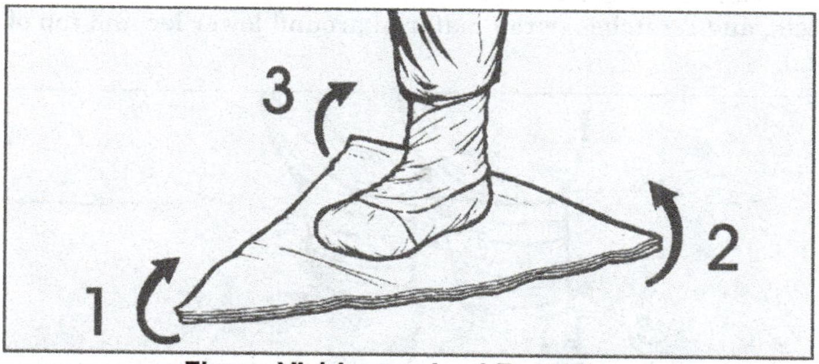

Figure VI-1 Improvised Foot Wear

(4) Fold front over the toes.

(5) Fold side corners, one at a time, over the instep.

(6) Secure by rope, vines, tape, etc., or tuck into other layers of material.

3. Other Protective Equipment

a. Sleeping bag.

(1) Fluff before use, ***especially*** at foot of bag.

(2) Air and dry daily to remove body moisture.

(3) Improvise with available material, dry grass, leaves, dry moss, etc.

b. Sun and snow goggles **(Figure VI-2)**.

(1) Wear in bright sun or snow conditions.

(2) Improvise by cutting small horizontal slits in webbing, bark, or similar materials.

Figure VI-2. Sun and Snow Goggles

c. Gaiters **(Figure VI-3)**. Used to protect from sand, snow, insects, and scratches (wrap material around lower leg and top of boots).

Figure VI-3. Gaiters

4. Shelters

Evasion considerations apply.

a. Site selection.

(1) Near signal and recovery site.

(2) Available food and water.

(3) Avoid natural hazards:

(a) Dead standing trees.
(b) Drainage and dry river beds except in combat areas.
(c) Avalanche areas.

(4) Location large and level enough to lie down in.

b. Types.

(1) Immediate shelters. Find shelter needing minimal improvements **(Figure VI-4)**.

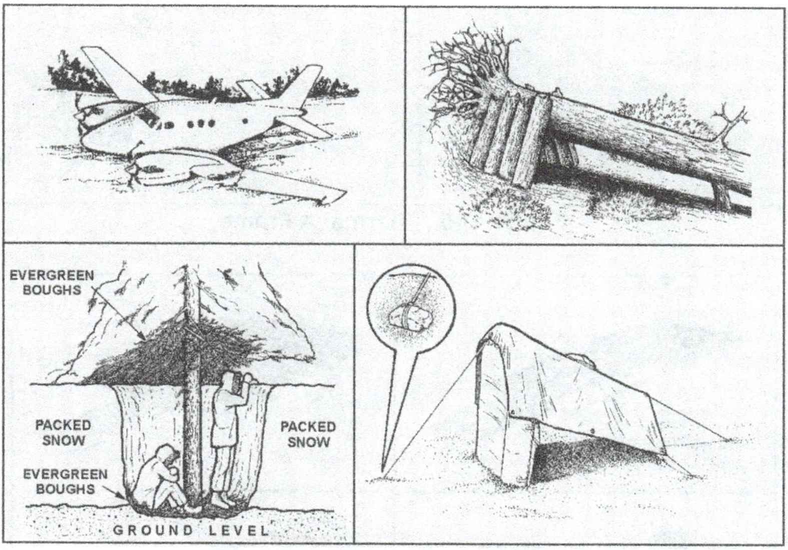

Figure VI-4. Immediate Shelters

(2) General shelter. Temperate climates require any shelter that gives protection from wind and rain.

(3) Thermal A Frame, Snow Trench, Snow Cave. **(Figures VI-5 through VI-7)**. Cold climates require an enclosed, insulated shelter.

(a) Snow is the most abundant insulating material.

(b) Air vent is required to prevent carbon monoxide poisoning when using an open flame inside enclosed shelters.

Note: As a general rule, unless you can see your breath, your snow shelter is too warm and should be cooled down to preclude melting and dripping.

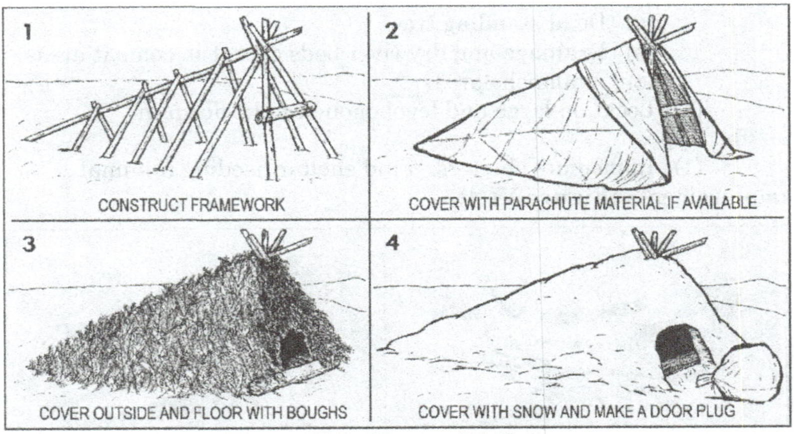

Figure VI-5. Thermal A Frame

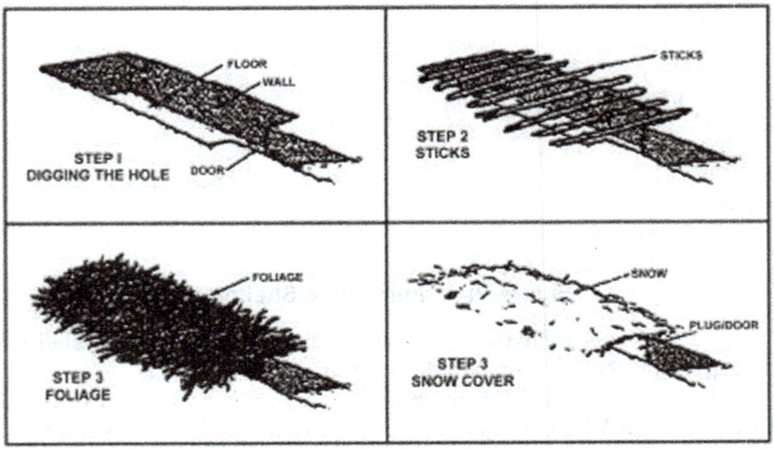

Figure VI-6. Snow Trench

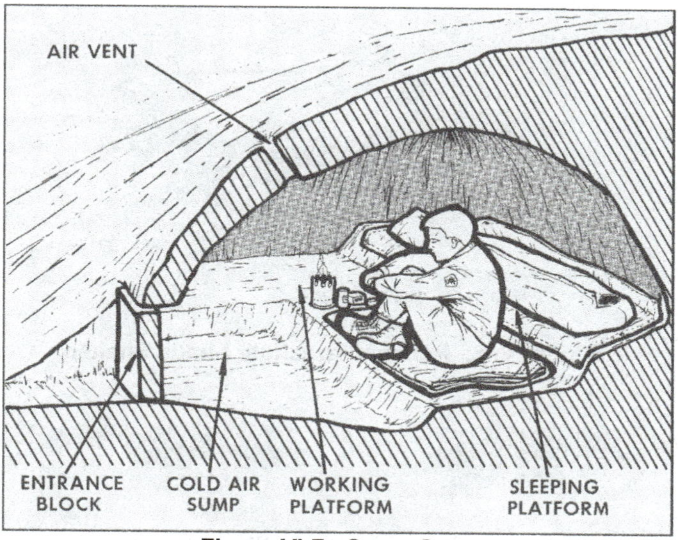

Figure VI-7. Snow Cave

(4) Shade shelter. Hot climates require a shade shelter to protect from ultraviolet rays **(Figure VI-8)**.

(a) To reduce the surface temperature, the shelter floor should be elevated or dug down (approximately 18 inches).

(b) For thermal protection, a minimum of **2** layers of material suspended 12-18 inches above the head is required. White is the best color to reflect heat (inner most layer should be of darker material).

(5) Elevated platform shelter **(Figure VI-9)**. Tropical/wet climates require enclosed, elevated shelter for protection from dampness and insects.

c. Shelter construction.

(1) Have entrance 45-90 degrees from prevailing wind.

(2) Cover with available material.

(a) If natural materials are used, arrange them in layers starting at the bottom with each layer overlapping the previous one. See **Figure VI-10** for an example.

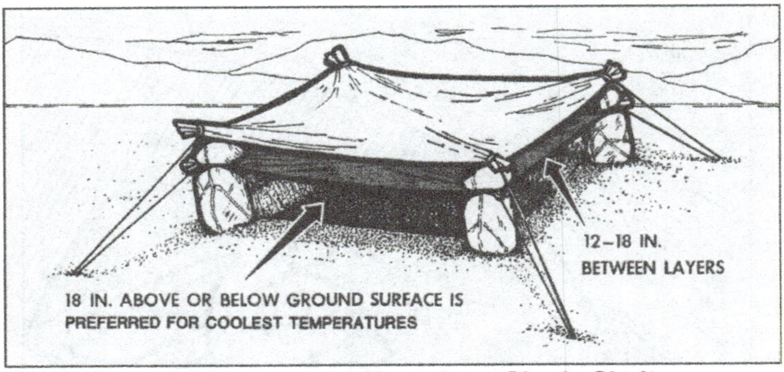

Figure VI-8. Poncho/Parachute Shade Shelter

Figure VI-9. Elevated Platform Shelter

Figure VI-10. Shingle Method

(b) If using porous material like parachute, blankets, etc.—

- Stretch as tight as possible
- Use a 40–60 degree slope.
- Use additional layers in heavy rain.

d. Shelter construction materials:

(1) Raft and raft parts.
(2) Vehicle or aircraft parts.
(3) Blankets, poncho, or parachute material.
(4) Sheet of plastic or plastic bag.
(5) Bark peeled off dead trees.
(6) Boughs, broad leaves, dry moss.
(7) Grass and sod.
(8) Snow.
(9) Sand and rocks.

e. Bed construction. Construct a bed to protect from cold, damp, ground using—

(1) Raft or foam rubber from vehicle seats.
(2) Boughs, leaves, or dry moss.

5. Fires

CAUTION: Weigh hazards and risks of detection against the need for a fire.

a. Evasion considerations:

(1) Use trees or other sources to dissipate smoke.
(2) Use fires at dusk, dawn, or during inclement weather.
(3) Use fires at times when the local populace is cooking.

b. Fire building. The **3** essential elements for starting a fire are heat, fuel, and oxygen.

(1) Heat sources:

(a) Matches or lighter.
(b) Flint and steel (experiment with various rocks and metals until a good spark is produced).
(c) Sparks from batteries.
(d) Concentrated sunlight (use magnifying glass or flashlight reflectors).
(e) Pyrotechnics, such as flares **(last resort)**, etc.

(f) Friction method **(Figure VI-11)**. Without prior training, this method is difficult to master and requires a lot of time to build the device.

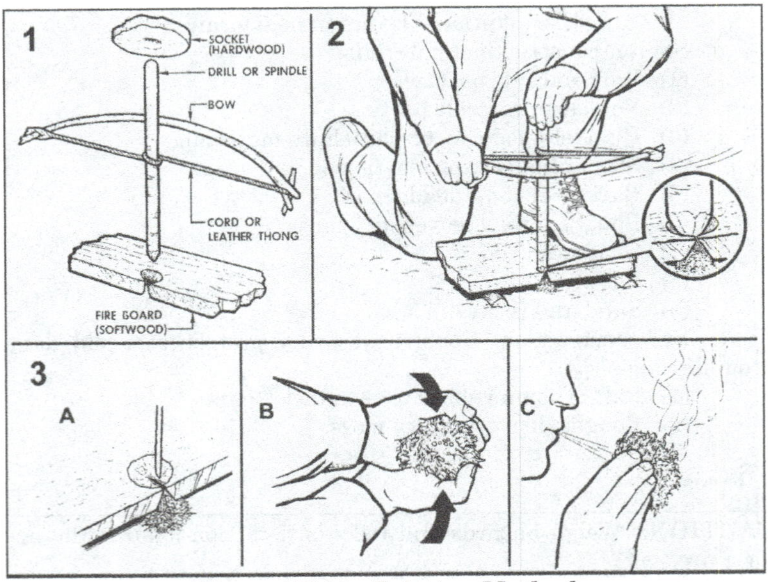

Figure VI-11. Friction Method

> **Note:** If possible, carry a fire-starting device with you.

(2) Fuel is divided into **3** categories: tinder, kindling, and fuel. (Gather large amounts of each category before igniting the fire.)

(a) **Tinder**. Tinder must be very finely shaved or shredded to provide a low combustion point and fluffed to allow oxygen to flow through. (To get tinder to burn hotter and longer, saturate with Vaseline, Chapstick, insect repellant, aircraft fuel, etc.) Examples of tinder include—

- Cotton.
- Candle (shred the wick, not the wax).
- Plastic spoon, fork, or knife.
- Foam rubber.

•Dry bark.
•Dry grasses.
•Gun powder.
•Pitch.
•Petroleum products.

(b) **Kindling**. Kindling must be small enough to ignite from the small flame of the tinder. Gradually add larger kindling until arriving at the size of fuel to burn.

(c) **Fuel**. Examples of fuel include—

•Dry hardwood (removing bark reduces smoke).
•Bamboo (open chambers to prevent explosion).
•Dry dung.

c. Types. Fires are built to meet specific needs or uses.

(1) Tepee fire **(Figure VI-12)**. Use the tepee fire to produce a concentrated heat source for cooking, lighting, or signaling.

Figure VI-12. Tepee Fire

(2) Log cabin fire **(Figure VI-13)**. Use the log cabin fire to produce large amounts of light and heat, to dry out wet wood, and provide coals for cooking, etc.

Figure VI-13. Log Cabin or Pyramid Fires

(3) Sod fire and reflector (**Figure VI-14**). Use fire reflectors to get the most warmth from a fire. Build fires against rocks or logs.

CAUTION: ***DO NOT*** use porous rocks or riverbed rock—they may explode when heated.

Figure VI-14. Sod Fire and Reflector

(4) Dakota fire hole (Figure VI-15). Use the Dakota fire hole for high winds or evasion situations.

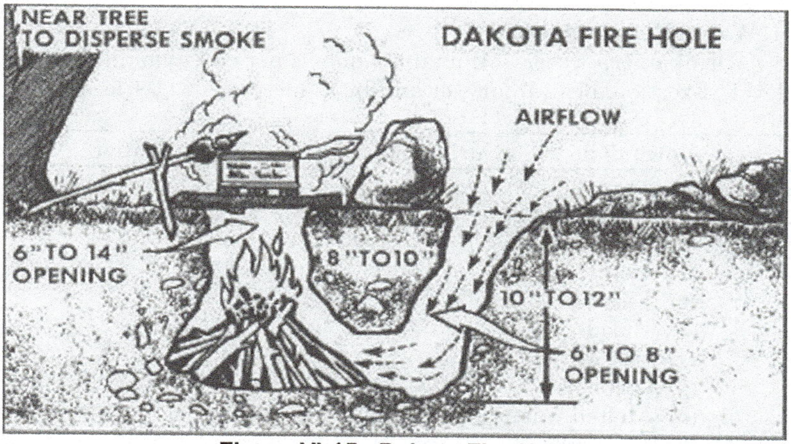

Figure VI-15. Dakota Fire Hole

(5) Improvised stoves **(Figure VI-16)**. These are very efficient.

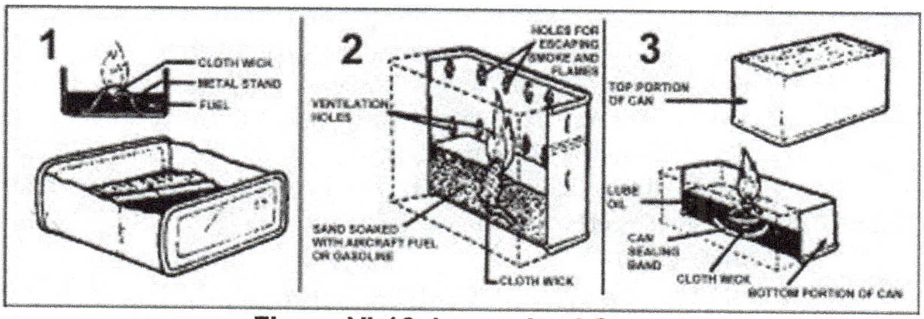

Figure VI-16. Improvised Stove

Chapter VII
Water

1. Water Requirements

Drink **extra** water. Minimum 2 quarts per day to maintain fluid level. Exertion, heat, injury, or an illness increases water loss.

> **Note:** Pale yellow urine indicates adequate hydration.

2. Water Procurement

a. ***DO NOT*** drink—

(1) Urine.
(2) Fish juices.
(3) Blood.
(4) Sea water.
(5) Alcohol.
(6) Melted water from new sea ice.

b. Water sources:

(1) Surface water (streams, lakes, and springs).
(2) Precipitation (rain, snow, dew, sleet) **(FigureVII-1)**.
(3) Subsurface (wells and cisterns).
(4) Ground water (when no surface water is available) **(Figure VII-2)**.

(a) Abundance of lush green vegetation.
(b) Drainages and low-lying areas.
(c) ***"V"*** intersecting game trails often point to water.
(d) Presence of swarming insects indicates water is near.
(e) Bird flight in the early morning or late afternoon might indicate the direction to water.

(5) Snow or ice.

(a) ***DO NOT*** eat ice or snow.

- Lowers body temperature.
- Induces dehydration.
- Causes minor cold injury to lips and mouth.

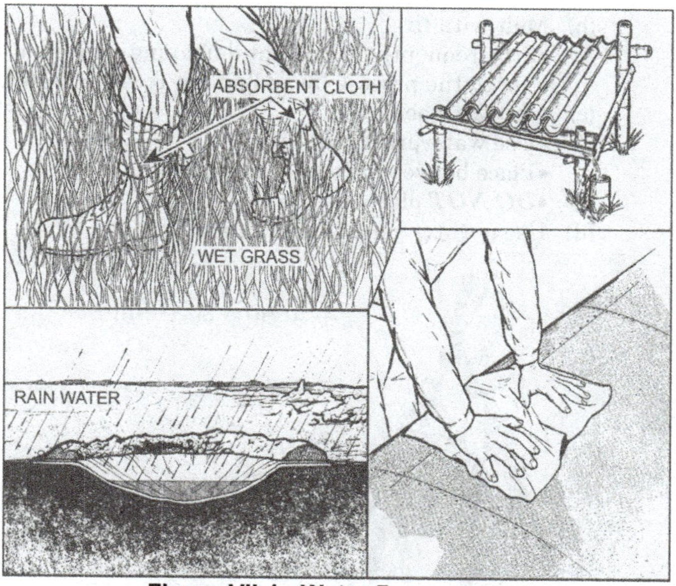

Figure VII-1. Water Procurement

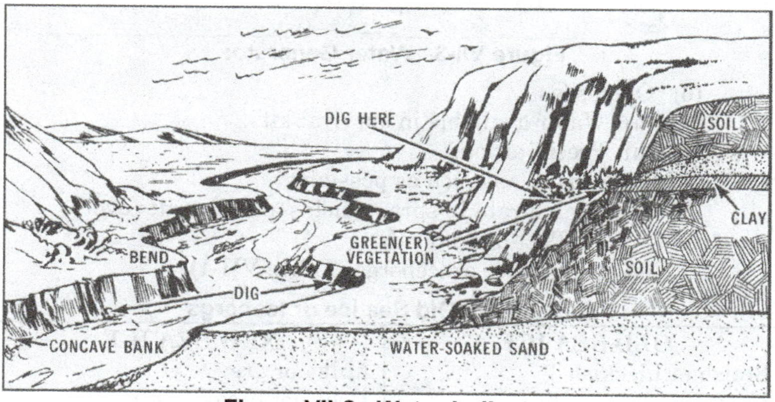

Figure VII-2. Water Indicators

(b) Melt with fire.
- •Stir frequently to prevent damaging container.
- •Speed the process by adding hot rocks or water.

(c) Melt with body heat.
- •Use waterproof container.
- •Place between layers of clothing.
- •***DO NOT*** place next to the skin.

(d) Use a water generator **(Figure VII-3)**.

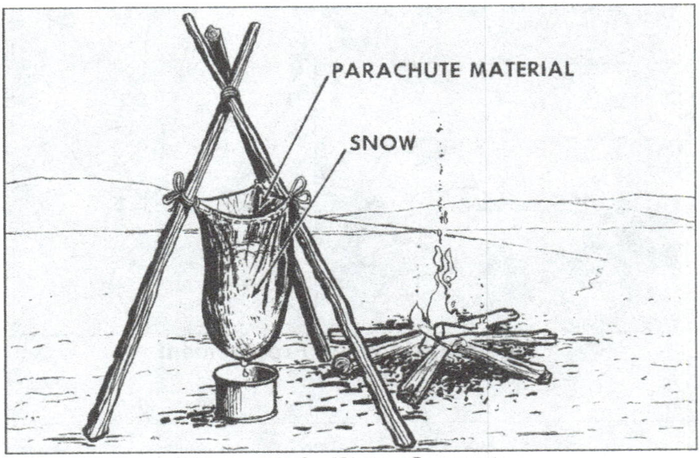

Figure VII-3. Water Generator

(6) Open seas.

(a) Water available in survival kits.

(b) Precipitation.
- •Drink as much as possible.
- •Catch rain in spray shields and life raft covers.
- •Collect dew off raft.

(c) Old sea ice or icebergs **(Table VII-1)**.

Table VII-1. Old Sea Ice or Icebergs

OLD SEA ICE	NEW SEA ICE
Bluish or blackish	Milky or grey
Shatters easily	Does not break easily
Rounded corners	Sharp edges
Tastes relatively salt-free	Tastes extremely salty

(7) Tropical areas.

(a) All open sources previously mentioned.

(b) Vegetation.

•Plants with hollow sections can collect moisture.

•Leaning Tree. Cloth absorbs rain running down tree and drips into container **(Figure VII-4)**.

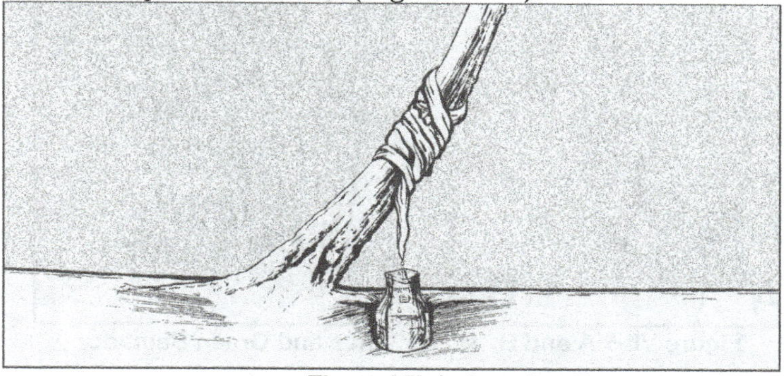

Figure VII-4. Leaning Tree

•Banana plants.

•Water trees (avoid milky sap).

••Tap before dark. Let sap stop running and harden during the daytime.

••Produce most water at night.

••For evasion situations, bore into the roots and collect water.

•Vines **(Figure VII-5A)**.

••Cut bark (***DO NOT*** use milky sap).

••If juice is clear and water like, cut as large a piece of vine as possible (cut the top first).

••Pour into hand to check smell, color, and taste to determine if drinkable.

••***DO NOT*** touch vine to lips.

••When water flow stops, cut off 6 inches of opposite end, water will flow again.

•Old bamboo.

••Shake and listen for water.

••Bore hole at bottom of section to obtain water.

••Cut out entire section to carry with you.

••Filter and purify.

•Green bamboo **(Figure VII-5B)**.

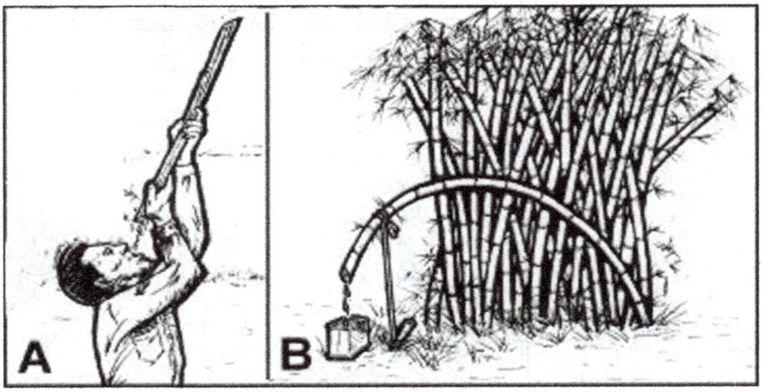

Figure VII-5 A and B. Water Vines and Green Bamboo

CAUTION: Liquid contained in green coconuts (**ripe** coconuts may cause diarrhea).

•Beach well. Along the coast, obtain water by digging a beach well **(Figure VII-6)**.

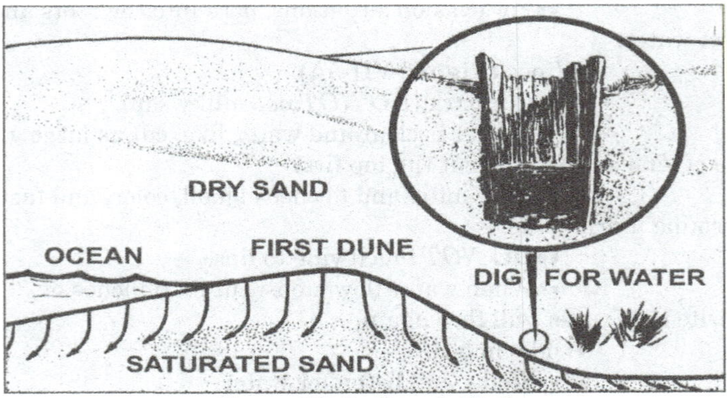

Figure VII-6. Beach Well

(8) Dry areas.
   (a) Solar still **(Figure VII-7)**.
   (b) Vegetation bag **(Figure VII-8)**.

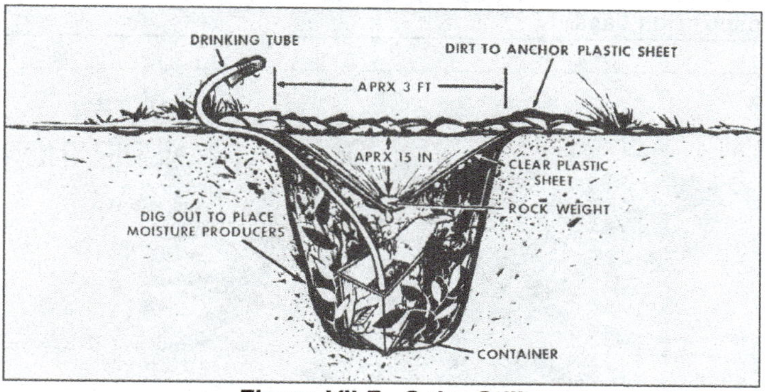

Figure VII-7. Solar Still

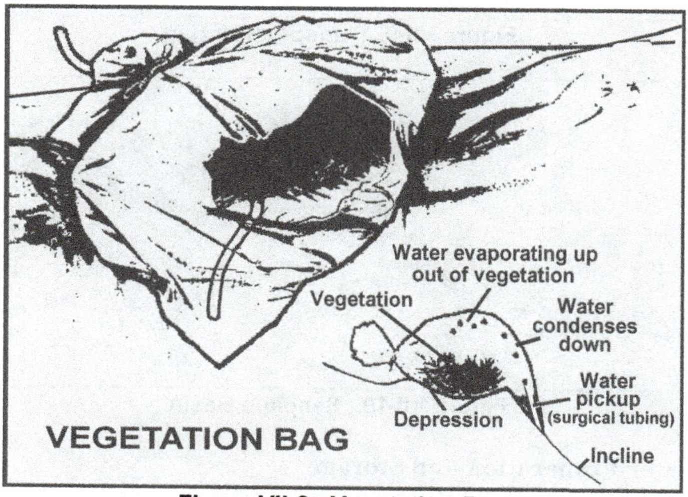

Figure VII-8. Vegetation Bag

   (c) Transpiration bag **(Figure VII-9)**.
   •Water bag must be clear.
   •Water will taste like the plant smells.

(d) Seepage basin **(Figure VII-10)**.

CAUTION: ***DO NOT*** use poisonous/toxic plants in vegetation/transpiration bags.

Figure VII-9. Transpiration Bag

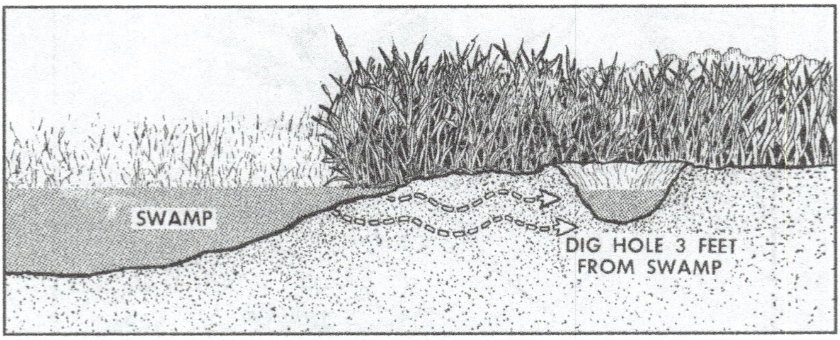

Figure VII-10. Seepage Basin

3. Water Preparation and Storage

a. Filtration. Filter through porous material (sand/charcoal).

b. Purification.

(1) Water from live plants requires no further treatment.

(2) Purify all other water.

(a) Boil at least 1 minute.

(b) Pour from one container to another to improve taste to aerate.

(c) Water purification tablets. Follow instructions on package.

c. Potable Water.

(1) If water cannot be purified, obtain water from a clear, cold, clean, and fast running source (if possible).

(2) Put in clear container and expose to the sun's ultraviolet rays to kill bacteria.

d. Storage. To prevent contamination, use a clean, covered or sealed container.

(1) Trash bag.
(2) Prophylactic.
(3) Section of bamboo.
(4) Flotation gear.

Chapter VIII

FOOD

1. Food Procurement

a. Sources and location.

(1) Mammals can be found where—

(a) Trails lead to watering, feeding, and bedding areas.

(b) Droppings or tracks look fresh.

(2) Birds can be found by—

(a) Observing the direction of flight in the early morning and late afternoon (leads to feeding, watering, and roosting areas).

(b) Listening for bird noises (indication of nesting areas

(3) Fish and other marine life locations **(Figure VIII-1)**.

Figure VIII-1. Fishing Locations

(4) Reptiles and amphibians are found almost worldwide.

(5) Insects are found—

(a) In dead logs and stumps.

(b) At ant and termite mounds.

(c) On ponds, lakes, and slow moving streams.

b. Procurement techniques.

(1) Snares—

(a) Work while unattended.

(b) Location:

•Trails leading to water, feeding, and bedding areas.

•Mouth of dens **(Figure VIII-2)**.

Figure VIII-2. Snare Placement

(c) Construction of simple loop snare.

•Use materials that will not break under the strain of holding an animal.

•Use a figure 8 (locking loop) if wire is used **(Figure VIII-3)**.

••Once tightened, the wire locks in place, preventing reopening, and the animal's escape.

•To construct a squirrel pole **(Figure VIII-4)** use simple loop snares.

•Make noose opening slightly larger than the animal's head **(3-finger** width for squirrels, **fist-sized** for rabbits**)**.

(d) Placement of snares (set as many as possible).

•Avoid disturbing the area.

•Use funneling (natural or improvised) **(Figure VIII-5)**.

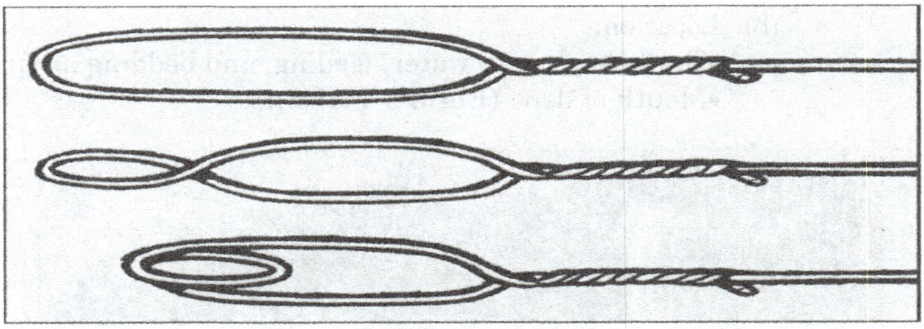

Figure VIII-3. Locking Loop

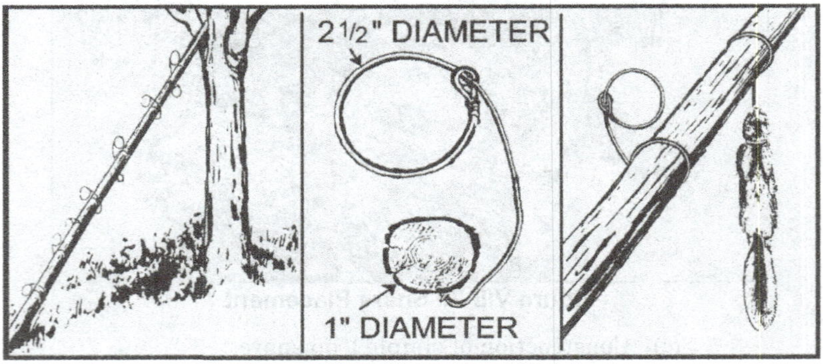

Figure VIII-4. Squirrel Pole

Figure VIII-5. Funneling

(2) Noose stick (easier and safer to use than the hands).

(3) Twist stick **(Figure VIII-6)**.

(a) Insert forked stick into a den until something soft is met.

(b) Twist the stick, binding the animal's hide in the fork.

(c) Remove the animal from the den.

(d) Be ready to **kill** the animal; **it may be dangerous**.

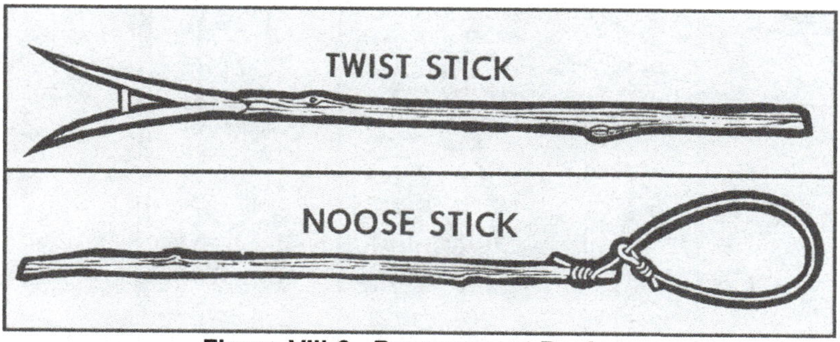

Figure VIII-6. Procurement Devices

(4) Hunting and fishing devices. (See **Figure VIII-7** for fishing procurement methods.)

(a) Club or rock.

(b) Spear.

(c) Slingshot.

(d) Pole, line, and hook.

(e) Net.

(f) Trap.

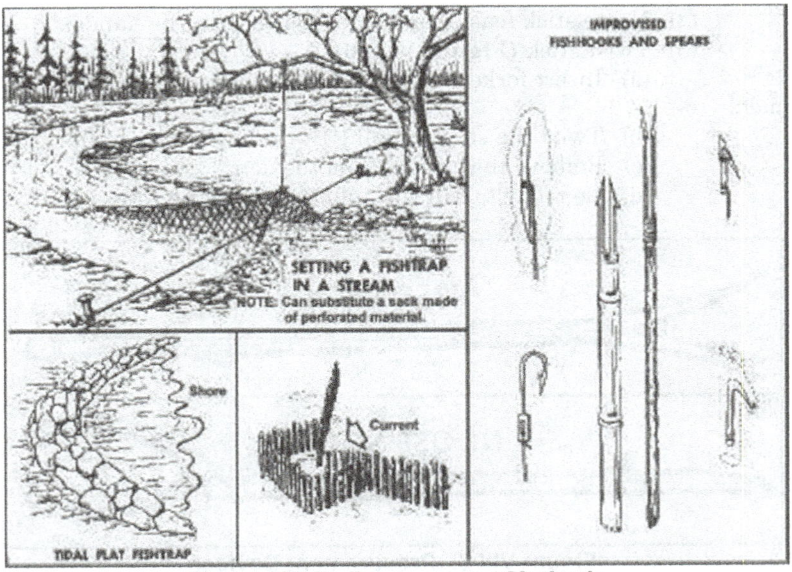

Figure VIII-7. Procurement Methods

(5) Precautions:

(a) Wear shoes to protect the feet when wading in water.

(b) Avoid reaching into dark holes.

(c) **Kill** animals before handling. Animals in distress may attract the enemy.

(d) ***DO NOT*** secure fishing lines to yourself or the raft.

(e) **Kill** fish before bringing them into the raft.

(f) ***DO NOT*** eat fish with—

- Spines.
- Unpleasant odor.
- Pale, slimy gills.
- Sunken eyes.
- Flabby skin.
- Flesh that remains dented when pressed.

(g) ***DO NOT*** eat fish eggs or liver (entrails).

(h) Avoid all crustaceans above the high tide mark.

(i) Avoid cone-shaped shells **(Figure VIII-8)**.

Figure VIII-8. Cone-Shaped Shells of Venomous Snails

(j) Avoid hairy insects; the hairs could cause irritation or infection.

(k) Avoid poisonous insects, for example:
- Centipedes.
- Scorpions.
- Poisonous spiders.

(l) Avoid disease carrying insects, such as—
- Flies.
- Mosquitoes.
- Ticks.

c. Plant Foods. ***Before using the following guide use your evasion chart to identify edible plants:***

> **Note**: If you cannot positively identify an edible plant and choose to try an unknown plant, these guidelines may help determine edibility.

(1) Selection criteria.

(a) Before testing for edibility, ensure there are enough plants to make testing worth your time and effort. Each part of a plant (roots, leaves, stems, bark, etc.) requires more than 24 hours to test. ***DO NOT*** waste time testing a plant that is not abundant.

(b) Test only **1** part of **1** plant at a time.

(c) Remember that eating large portions of plant food on an empty stomach may cause diarrhea, nausea, or cramps. ***Two*** good examples are ***green apples*** and ***wild onions***. Even after testing food and finding it safe, eat in moderation.

(2) Avoid plants with the following characteristics:

> **Note:** Using these guidelines in selecting plants for food may eliminate some edible plants; however, these guidelines will help prevent choosing potentially toxic plants.

(a) Milky sap (dandelion has milky sap but is safe to eat and easily recognizable).

(b) Spines, fine hairs, and thorns (skin irritants/contact dermatitis). ***Prickly pear*** and ***thistles*** are exceptions. ***Bracken fern fiddleheads*** also violate this guideline.

(c) Mushrooms and fungus.

(d) Umbrella shaped flowers (hemlock is eliminated).

(e) Bulbs (***only*** onions smell like onions).

(f) Grain heads with pink, purplish, or black spurs.

(g) Beans, bulbs, or seeds inside pods.

(h) Old or wilted leaves.

(i) Plants with shiny leaves.

(j) White and yellow berries. (Aggregate berries such as black and dewberries are always edible, test all others before eating.)

(k) Almond scent in woody parts and leaves.

d. Test procedures.

CAUTION: Test all parts of the plant for edibility. Some plants have both edible and inedible parts. **NEVER ASSUME** a part that proved edible when cooked is edible raw, test the part raw before eating. The same part or plant may produce varying reactions in different individuals.

(1) Test only **1** part of a plant at a time.

(2) Separate the plant into its basic components (stems, roots, buds, and flowers).

(3) Smell the food for strong acid odors. Remember, smell alone does not indicate a plant is edible or inedible.

(4) ***DO NOT*** eat 8 hours before the test and drink only purified water.

(5) During the 8 hours you abstain from eating, test for contact poisoning by placing a piece of the plant on the inside of your elbow or wrist. The sap or juice should contact the skin. Usually 15 minutes is enough time to allow for a reaction.

(6) During testing, take ***NOTHING*** by mouth **EXCEPT** purified water and the plant you are testing.

(7) Select a small portion of a single part and prepare it the way you plan to eat it.

(8) Before placing the prepared plant in your mouth, touch a small portion (a pinch) to the outer surface of your lip to test for burning or itching.

(9) If after 3 minutes there is no reaction on your lip, place the plant on your tongue and hold it for 15 minutes.

(10) If there is no reaction, thoroughly chew a pinch and hold it in your mouth for 15 minutes **(*DO NOT* SWALLOW)**. If any ill effects occur, rinse out your mouth with water.

(11) If nothing abnormal occurs, swallow the food and wait 8 hours. If **any ill effects** occur during this period, **induce** vomiting and drink a water and charcoal mixture.

(12) If no ill effects occur, eat **¼ cup** of the same plant prepared the same way. Wait another 8 hours. If no ill effects occur, the plant part as prepared is safe for eating.

CAUTION:
1. Ripe tropical fruits should be peeled and eaten raw. Softness, rather than color, is the best indicator of ripeness. Cook unripe fruits and discard seeds and skin.
2. Cook underground portions when possible to reduce bacterial contamination and ease digestion of their generally high starch content.
3. During evasion, you may not be able to cook. Concentrate your efforts on leafy green plants, ripe fruits, and above ground ripe vegetables not requiring significant preparation.

2. Food Preparation

Animal food gives the greatest food value per pound.

a. Butchering and skinning.

(1) Mammals.

(a) Remove the skin and save for other uses.

(a) One cut skinning of small game **(Figure VIII-9)**.

•Open the abdominal cavity.

•Avoid rupturing the intestines.

•Remove the intestines.

•Save inner organs (heart, liver, and kidneys) and all meaty parts of the skull, brain, tongue, and eyes.

(b) Wash when ready to use.

(c) If preserving the meat, remove it from the bones.

(d) Unused or inedible organs and entrails may be used as bait for other game.

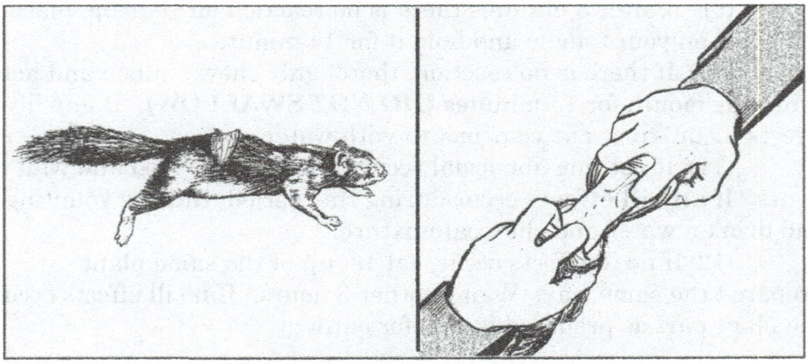

Figure VIII-9. Small Game Skinning

(2) Frogs and snakes.

(a) Skin.

(b) Discard skin, head with 2 inches of body, and internal organs.

(3) Fish.

(a) Scale (if necessary) and gut fish soon after it is caught.

(b) Insert knifepoint into anus of fish and cut open the belly.

(c) Remove entrails.

(d) Remove gills to prevent spoilage.

(4) Birds.

(a) Gut soon after killing.

(b) Protect from flies.

(c) Skin or pluck them.

(d) Skin scavengers and sea birds.

(5) Insects.

(a) Remove all hard portions such as the legs of grasshoppers or crickets. (The rest is edible.)

(b) Recommend cooking grasshopper-size insects.

CAUTION: Dead insects spoil rapidly, ***DO NOT*** save.

(6) Fruits, berries, and most nuts can be eaten raw.

b. Cooking.

CAUTION: To kill parasites, thoroughly cook all wild game, freshwater fish, clams, mussels, snails, crawfish, and scavenger birds. Saltwater fish may be eaten raw.

(1) Boiling (most nutritious method of cooking—drink the broth).

(a) Make metal cooking containers from ration cans.

(b) Drop heated rocks into containers to boil water or cook food.

(2) Baking.

(a) Wrap in leaves or pack in mud.

(b) Bury food in dirt under coals of fire.

(3) Leaching. Some nuts (acorns) must be leached to remove the bitter taste of tannin. Use one of the following leaching methods:

(a) First method:

•Soaking and pouring the water off.

•Crushing and pouring water through. Cold water should be tried first; however, boiling water is sometimes best.

•Discarding water.

(b) Second method:

•Boil, pour off water, and taste the plant.

•If bitter, repeat process until palatable.

(4) Roasting.

(a) Shake shelled nuts in a container with hot coals.

(b) Roast thinly sliced meat and insects over a candle.

3. Food Preservation

b. Keeping an animal alive.

c. Refrigerating.

(1) Long term.

(a) Food buried in snow maintains a temperature of approximately 32 degrees F.

(b) Frozen food will not decompose (freeze in meal-size portions).

(2) Short term.

(a) Food wrapped in waterproof material and placed in a stream remains cool in summer months.

(b) Earth below the surface, particularly in shady areas or along streams, is cooler than the surface.

(c) Wrap food in absorbent material such as cotton and re-wet as the water evaporates.

c. Drying and smoking removes moisture and preserves food.

(1) Use salt to improve flavor and promote drying.

(2) Cut or pound meat into thin strips.

(3) Remove fat.

(4) ***DO NOT*** use pitch woods such as fir or pine; they produce soot giving the meat an undesirable taste.

d. Protecting meat from animals and insects.

(1) Wrapping food.

(a) Use clean material.

(b) Wrap pieces individually.

(c) Ensure all corners of the wrapping are insect proof.

(d) Wrap soft fruits and berries in leaves or moss.

(2) Hanging meat.

(a) Hang meat in the shade.

(b) Cover during daylight hours to protect from insects.

(3) Packing meat on the trail.

(a) Wrap before flies appear in the morning.

(b) Place meat in fabric or clothing for insulation.

(c) Place meat inside the pack for carrying. Soft material acts as insulation helping keep the meat cool.

(d) Carry shellfish, crabs, and shrimp in wet seaweed.

e. ***DO NOT*** store food in the shelter; it attracts unwanted animals.

Chapter IX

INDUCED CONDITIONS

(NUCLEAR, BIOLOGICAL, AND CHEMICAL CONSIDERATIONS)

1. Nuclear Conditions

CAUTION: Radiation protection depends on time of exposure, distance from the source, and shielding.

a. Protection.

(1) **FIND PROTECTIVE SHELTER IMMEDIATELY!**

(2) Gather all equipment for survival (time permitting).

(3) Avoid detection and capture.

(a) Seek existing shelter that may be improved **(Figure IX-1)**.

Figure IX-1. Immediate Action Shelter

(b) If no shelter is available, dig a trench or foxhole as follows:

•Dig trench deep enough for protection, then enlarge for comfort **(Figure IX-2)**.

•Cover with available material.

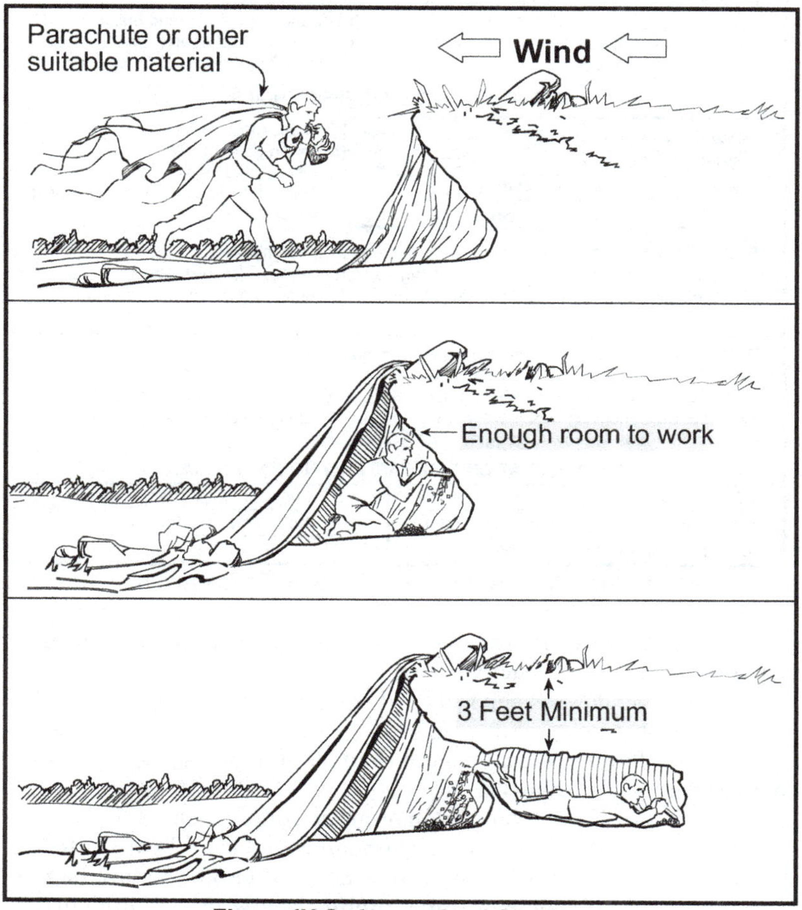

Figure IX-2. Improvised Shelter

(4) Radiation shielding efficiencies **(Figure IX-3)**.

NUCLEAR EXPLOSIONS: Fall flat. Cover exposed body parts. Present minimal profile to direction of blast. ***DO NOT** look at fireball!* Remain prone until blast effects are over.

SHELTER: Pick, as soon as possible, 5 minutes unsheltered is maximum!
Priority:
(1) Cave or tunnel covered with 3 or more feet of earth.
(2) Storm/storage cellars
(3) Culverts.
(4) Basements.
(5) Abandoned stone/mud buildings.
(6) Foxhole 4 feet deep (remove topsoil within 2 feet radius of foxhole lip).

RADIATION SHIELDING EFFICIENCIES

Iron/Steel	.7 inches	**Cinder Block**	5.3 inches
Brick	2.0 inches	**Ice**	6.8 inches
Concrete	2.2 inches	**Wood (Soft)**	8.8 inches
Earth	3.3 inches	**Snow**	20.3 inches

One thickness reduces received radiation dose by 1/2. Additional thickness added to any amount of thickness reduces received radiation dose by 1/2.

SHELTER SURVIVAL: Keep contaminated materials out of shelter.
Good Weather: Bury contaminated clothing outside of shelter (recover later).
Bad Weather: Shake strongly or beat with branches. Rinse and /or shake wet clothing. ***DO NOT** wring out!*

PERSONAL HYGIENE: Wash entire body with soap and *any* water; give close attention to fingernails and hairy parts.
No Water: Wipe all exposed skin surfaces with clean cloth or uncontaminated soil. Fallout/dusty conditions--keep entire body covered. Keep handkerchief/cloth over mouth and nose. Improvise goggles. ***DO NOT** smoke!*

DAILY RADIATION TIME TABLE for NO RATE METER

4-6	Complete isolation	**9-12**	2-4 hours exposure per day
3-7	Brief exposure (30 minutes maximum)	**13**	Normal movement.
8	Brief exposure (1 hour maximum)		

Figure IX-3. Radiation Shielding Efficiencies

(5) Leave contaminated equipment and clothing near shelter for retrieval after radioactive decay.

(6) Lie down, keep warm, sleep, and rest.

b. Substance:

(1) Water. Allow no more than 30 minutes exposure on **3d** day for water procurement.

(a) Water sources (in order of preference):

- Springs, wells, or underground sources are **safest.**
- Water in pipes/containers in abandoned buildings.
- Snow (**6** or **more inches below** the surface during the fallout).

•Streams and rivers (filtered before drinking).

•Lakes, ponds, pools, etc.

•Water from below the surface (***DO NOT*** stir up the water).

•Use a seep well.

(b) Water preparation **(Figures IX-4 and IX-5)**.

•Filtering through earth removes 99 percent of radioactivity.

•Purify all water sources.

(2) Food.

(a) Processed foods (canned or packaged) are preferred; wash and wipe containers before use.

(b) Animal foods.

•Avoid animals that appear to be sick or dying.

•Skin carefully to avoid contaminating the meat.

•Before cooking, cut meat away from the bone, leaving at least 1/8 inch of meat on the bone.

•Discard all internal organs.

•Cook all meat until **very well** done.

(c) Avoid.

•Aquatic food sources (use only in extreme emergencies because of high concentration of radiation).

•Shells of all eggs (contents will be safe to eat).

•Milk from animals.

(d) Plant foods (in order of preference).

•Plants whose edible portions grow underground (for example, potatoes, turnips, carrots, etc.). Wash and remove skin.

•Edible portions growing above ground that can be washed and peeled or skinned (bananas, apples, etc.).

•Smooth skinned vegetables, fruits, or above ground plants that are not easily peeled or washed.

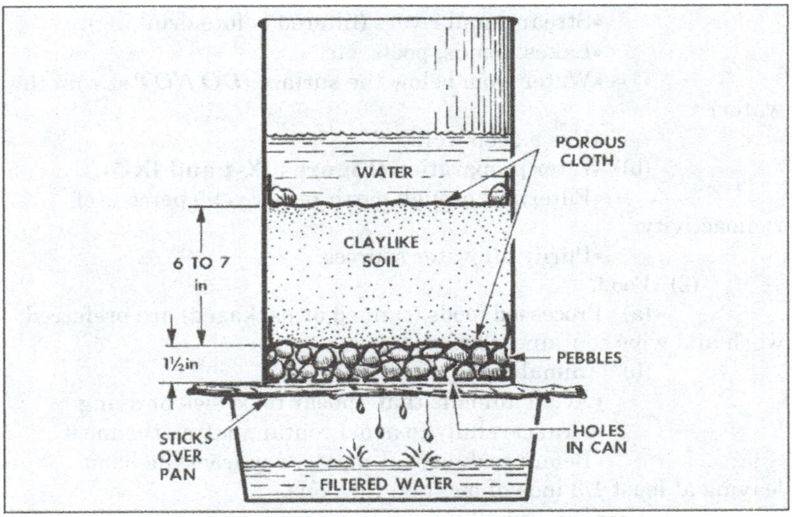

Figure IX-4. Filtration Systems, Filtering Water

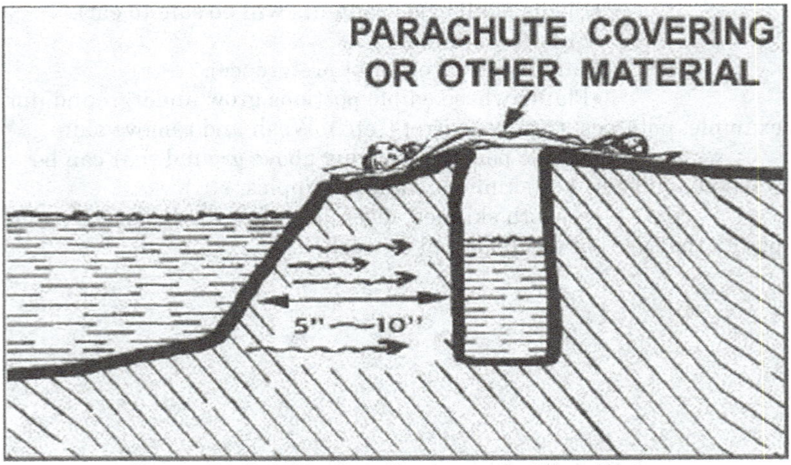

Figure IX-5. Filtration Systems, Settling Water

c. Self-aid:

(1) General rules:

(a) Prevent exposure to contaminants.

(b) Use personal hygiene practices and remove body waste from shelter.

(c) Rest, avoid fatigue.

(d) Drink liquids.

(2) Wounds.

(a) Clean affected area.

(b) Use antibacterial ointment or cleaning solution.

(c) Cover with clean dressing.

(d) Watch for signs of infection.

(3) Burns.

(a) Clean affected area.

(b) Cover with clean dressing.

(4) Radiation sickness (nausea, weakness, fatigue, vomiting, diarrhea, loss of hair, radiation burns).

(a) Time is required to overcome.

(b) Rest.

(c) Drink fluids.

(d) Maintain food intake.

(e) Prevent additional exposure.

2. Biological Conditions

a. Clues which may alert you to a biological attack follow:

(1) Enemy aircraft dropping objects or spraying.

(2) Breakable containers or unusual bombs, particularly those bursting with little or no blast, and muffled explosions.

(3) Smoke or mist of unknown origin.

(4) Unusual substances on the ground or vegetation; sick looking plants or crops.

b. Protection from biological agents follow:

(1) Use protective equipment.

(2) Bathe as soon as the situation permits.

(3) Wash hair and body thoroughly with soap and water.

(4) Clean thoroughly under fingernails.

(5) Clean teeth, gums, tongue, and roof of mouth frequently.

c. Survival tips for biological conditions follow:

(1) Keep your body and living area clean.

(2) Stay alert for clues of biological attack.

(3) Keep nose, mouth, and skin covered.

(4) Keep food and water protected. Bottled or canned foods are safe if sealed. If in doubt, boil food and water for 10 minutes.

(5) Construct shelter in a clear area, away from vegetation, with entrance 90 degrees to the prevailing wind.

(6) If traveling, travel crosswind or upwind (taking advantage of terrain to stay away from depressions).

3. Chemical Conditions

a. Detecting.

(1) Smell. Many agents have little or no odor.

(2) Sight. Many agents are colorless:

(a) Color. Yellow, orange, or red smoke or mist.

(b) Liquid. Oily, dark patches on leaves, ground, etc.

(c) Gas. Some agents appear as a mist immediately after shell burst.

(d) Solid. Most solid state agents have some color.

(3) Sound. Muffled explosions are possible indications of chemical agent bombs.

(4) Feel. Irritation to the nose, eyes, or skin and/or moisture on the skin are danger signs.

(5) Taste. Strange taste in food or water indicates contamination.

(6) General indications. Tears, difficult breathing, choking, itching, coughing, dizziness.

(7) Wildlife. Presence of sick or dying animals.

b. Protection against chemical agents follows:

(1) Use protective equipment.

(2) Avoid contaminated areas.

(a) Exit contaminated area by moving crosswind.

(b) Select routes on high ground.

(c) Avoid cellars, ditches, trenches, gullies, valleys, etc.

(d) Avoid woods, tall grasses, and bushes as they tend to hold chemical agent vapors.

(e) Decontaminate body and equipment as soon as possible by—

- •Removing. Pinch-blotting.
- •Neutralizing. Warm water.
- •Destroying. Burying.

c. Self-aid in chemically contaminated areas.

(1) If a chemical defense ensemble is available—

(a) Use all protective equipment.

(b) Follow antidote directions when needed.

(2) If a chemical defense ensemble is not available—

(a) Remove or tear away contaminated clothing.

(b) Rinse contaminated areas with water.

(c) Improvise a breathing filter using materials available (T-shirt, handkerchief, fabric, etc.).

d. Tips for the survivor:

(1) ***DO NOT*** use wood from a contaminated area for fire.

(2) Look for signs of chemical agents around water sources before procurement (oil spots, foreign odors, dead fish, or animals).

(3) Keep food and water protected.

(4) ***DO NOT*** use plants for food or water in contaminated areas.

Appendix A

THE WILL TO SURVIVE

ARTICLE VI CODE OF CONDUCT

I will never forget that I am an American fighting for freedom, responsible for my actions, and dedicated to the principles which made my country free. I will trust in my God and in the United States of America.

1. Psychology of Survival

a. Preparation—

(1) Know your capabilities and limitations.

(2) Keep a positive attitude.

(3) Develop a realistic plan.

(4) Anticipate fears.

(5) Combat psychological stress by—

(a) Recognizing and anticipating existing *stressors* (injury, death, fatigue, illness, environment, hunger, isolation).

(b) Attributing normal reactions to existing *stressors* (fear, anxiety, guilt, boredom, depression, anger).

(c) Identifying signals of distress created by *stressors* (indecision, withdrawal, forgetfulness, carelessness, and propensity to make mistakes).

b. Strengthen your will to survive with—

(1) The Code of Conduct.

(2) Pledge of Allegiance.

(3) Faith in America.

(4) Patriotic songs.

(5) Thoughts of return to family and friends.

c. Group dynamics of survival include—

(1) Leadership, good organization, and cohesiveness promote high morale:

(a) Preventing panic.

(b) Creating strength and trust in one another.

(c) Favoring persistency in overcoming failure.

(d) Facilitating formulation of group goals.

(2) Taking care of your buddy.

(3) Working as a team.

(4) Reassuring and encouraging each other.

(5) Influencing factors are—
   (a) Enforcing the chain of command.
   (b) Organizing according to individual capabilities.
   (c) Accepting suggestions and criticism.

2. Spiritual Considerations

a. Collect your thoughts and emotions.
b. Identify your personal beliefs.
c. Use self-control.
d. Meditate.
e. Remember past inner sources to help you overcome adversity.
f. Pray for your God's help, strength, wisdom, and rescue.
   (1) Talk to your God.
   (2) Give thanks that God is with you.
   (3) Ask for God's help.
   (4) Pray for protection and a positive outcome.
g. Remember scripture, verses, or hymns; repeat them to yourself and to your God.
h. Worship without aid of written scripture, clergy, or others.
i. Forgive—
   (1) Yourself for what you have done or said that was wrong.
   (2) Those who have failed you.
j. Praise God and give thanks because—
   (1) God is bigger than your circumstances.
   (2) God will see you through (no matter what happens).
   (3) Hope comes from a belief in heaven and/or an after-life.
k. Trust.
   (1) Faith and trust in your God.
   (2) Love for family and self.
   (3) Never lose hope.
   (4) Never give up.
l. With other survivors—
   (1) Identify or appoint a religious lay leader.
   (2) Discuss what is important to you.
   (3) Share scriptures and songs.
   (4) Pray for each other.
   (5) Try to have worship services.

(6) Write down scriptures and songs that you remember.

(7) Encourage each other while waiting for rescue, remember—

(a) Your God loves you.

(b) Praise your God.

Appendix B
Publication Information

1. Scope

This UNCLASSIFIED multiservice tactics, techniques, and procedures publication is designed to assist Service members in a survival situation regardless of geographic location.

2. Purpose

This publication provides Service members a quick reference, weatherproof, pocket-sized guide on basic survival, evasion, and recovery information.

3. Application

The target audience for this publication is any Service member requiring basic survival, evasion, and recovery information.

4. Implementation Plan

Participating Service command offices of primary responsibility (OPRs) will review this publication, validate the information, and reference and incorporate it in Service and command manuals, regulations, and curricula as follows:

Army. The Army will incorporate the procedures in this publication in US Army training and doctrinal publications as directed by the commander, US Army Training and Doctrine Command (TRADOC). Distribution is in accordance with the Department of Army Form 12-99-R.

Marine Corps. The Marine Corps will incorporate the procedures in this publication in US Marine Corps training and doctrinal publications as directed by the commanding general, US Marine Corps Combat Development Command (MCCDC). Distribution is in accordance with Marine Corps Publication Distribution System.

Navy. The Navy will incorporate these procedures in US Navy doctrinal and training publications as directed by the commander, Navy Warfare Development Command (NWDC). Distribution is in accordance with Military Standard Requisitioning and Issue Procedures Desk Guide and Navy Standing Operating Procedures Publication 409.

Air Force. Air Force units will validate and incorporate appropriate procedures in accordance with applicable governing directives. Distribution is in accordance with Air Force Instructions 33-360.

5. User Information

a. The TRADOC-MCCDC-NWDC-AFDC Air Land Sea Application (ALSA) Center developed this publication with the joint participation of the approving Service commands. ALSA will review and update this publication as necessary.

b. This publication reflects current joint and Service doctrine, command and control (C2) organizations, facilities, personnel, responsibilities, and procedures. Changes in Service protocol, appropriately reflected in joint and Service publications, will likewise be incorporated in revisions to this document.

c. We encourage recommended changes for improving this publication. Key your comments to the specific page and paragraph and provide a rationale for each recommendation. Send comments and recommendation directly to—

Army

Commander
US Army Training and Doctrine Command
ATTN: ATDO-A
Fort Monroe VA 23651-5000
DSN 680-3153 COMM (757) 727-3153

Marine Corps

Commanding General
US Marine Corps Combat Development Command
ATTN: C42
3300 Russell Road
Quantico VA 22134-5001
DSN 278-6234 COMM (703) 784-6234

Navy

Commander
Navy Warfare Development Command (Det Norfolk)
ATTN: ALSA Liaison Officer
1530 Gilbert Street
Norfolk VA 23511-2785
DSN 262-2782 COMM (757) 322-2782
E-mail: ndcjoint@nctamslant.navy.mil

Air Force

HQ Air Force Doctrine Center
ATTN: DJ
216 Sweeney Boulevard Suite 109
Langley AFB VA 23665-2722
DSN 754-8091 COMM (757) 764-8091
E-mail Address: afdc.dj@langley.af.mil

ALSA

ALSA Center
ATTN: Director
114 Andrews Street
Langley AFB, VA 23665-2785
DSN 575-0902 COMM (757) 225-0902
E-mail: alsadirector@langley.af.mil

TRAINING CIRCULAR
No. 21-3

HEADQUARTERS
DEPARTMENT OF THE ARMY
Washington, DC, 17 March 1986

SOLDIER'S HANDBOOK FOR INDIVIDUAL OPERATIONS AND SURVIVAL IN COLD-WEATHER AREAS

TABLE OF CONTENTS

*** This publication supersedes TC 21-3, 30 September 1974.**

CHAPTER 4. SMALL ARMS AND AMMUNITION

CHAPTER 5. RATIONS AND DIET

CHAPTER 6. HYGIENE AND FIRST AID

CHAPTER 7. BIVOUAC ROUTINE

CHAPTER 8. INDIVIDUAL MOVEMENT

CHAPTER 9. LAND NAVIGATION

CHAPTER 10. SURVIVAL

PREFACE

This handbook is an aid for soldiers in cold climates. However, expertise in dealing with cold-weather operations cannot be gained just from reading this handbook. To develop successful cold-weather field techniques, application training and operational experience are necessary. Additional information is available in FM 31-70, FM 31-71, FM 90-11 (TBP), and FM 9-207.

Military history of operations in northern areas shows that man's success or failure is measured by his regard for the dominant characteristics of a cold-weather environment. For a soldier to become a cold-weather expert, he must learn the techniques and rules needed to fight and live in a cold-weather climate.

The proponent of this publication is HQ TRADOC. Submit changes for improving this publication on DA Form 2028 (Recommended Changes to Publications and Blank Forms) and forward it to Commandant, US Army Infantry School, ATTN: ATSH-B, Fort Benning, Georgia 31905-5593.

Unless otherwise stated, whenever the masculine gender is used, both men and women are included.

CHAPTER 1

Introduction to Cold-Weather Operations

"Soldiers in pairs look after each other; be faithful to your mate; be loyal to your leaders."

Anonymous

1-1. COLD CONDITIONS

COLD-WET

Cold-wet conditions occur when temperatures are near freezing and high and low temperatures cause alternate freezing and thawing. These conditions can be accompanied by wet snow and rain, causing the ground to become slushy and muddy. Soldiers need clothing that has a waterproof or water-repellent, wind-resistant outer layer, and an inner layer with insulation that provide protection in moderately cold weather (above 14°F). **Waterproof footwear is essential.**

COLD-DRY

Cold-dry conditions occur when average temperatures are lower than 14°F. The ground is usually frozen and the snow dry. Low temperatures plus wind increase the need for protection of the entire body. For these conditions, soldiers need clothing that will provide insulation for a wind-chill factor of -80°F. For temperatures below 14°F, white vapor barrier (VB) boots effective down to temperatures of -50°F must be worn.

WIND CHILL

Wind chill is a measure of the combined effects of wind and temperature. Temperature alone does not give a true indication of the impact of the cold. In order to effectively gauge the difference between temperature and the impact of the cold, a wind-chill scale must be used. The chart in Appendix A is a simple and practical guide showing the conditions under which cold weather is dangerous and when exposed flesh is likely to freeze. Soldiers must be familiar with the wind-chill chart to recognize these dangers.

1-2. EFFECTS OF COLD WEATHER ON MILITARY OPERATIONS

Cold has a numbing effect on skin, which hinders task performance. This must be considered when accomplishing such routine tasks as vehicle maintenance, and making and striking bivouac areas (FM 9-207).

When conditions become extreme and the temperature drops, the problem of survival becomes critical. Under these circumstances soldiers may tend to withdraw emotionally, not wanting to leave the warmth and protection of their shelter or tent. The remedy is physical activity. Try to remain alert and active. Execution of all military duties and the fast and proper performance of group tasks are vital.

COCOON-LIKE EXISTENCE

Many soldiers, when wearing several layers of clothing and with their heads covered, tend to withdraw within themselves and assume a cocoon-like existence. When so clothed, many soldiers' hearing and field of vision are restricted; they tend not to notice their surroundings. Their thinking and reasoning become sluggish.

INDIVIDUAL AND GROUP HIBERNATION

This process is characterized by individuals seeking the comfort of sleeping bags, and by the group remaining in tents or other shelter. Guard and security measures may be neglected, jeopardizing the security and safety of the unit.

1-3. OVERCOMING THE COLD

The body must be protected. It must be kept clean, dry, and warm, with normal body processes maintained. Rest and nourishment are vital. The right approach to cold-weather living will keep you healthy and get you through challenging times. There are four basic rules to follow:

KEEP IN SHAPE

Cold-weather clothing and equipment are heavy, and add weight to your normal equipment load. This hinders movement in snow and uses too much energy. The importance of being in excellent physical condition cannot be overemphasized. Your unit must have a tough, challenging, demanding physical training program.

DRINK PLENTY OF WATER

Water may be hard to get, so you may drink less than you need or only when you are thirsty. This will not give you the water needed to avoid dehydration. Drink plenty of water (at least 3.5 quarts per day) to avoid dehydration and fatigue. DO NOT eat snow as a water substitute; it will lower your body's core temperature.

EAT TO KEEP FIT

Regular, nutritious hot food is needed for top performance. You must eat, even when you are not hungry.

KEEP A GOOD ATTITUDE

You will meet many new challenges, but none that cannot be overcome. Your attitude toward the cold will reflect those of your leaders.

1-4. POSITIVE LEADERSHIP AND THE RIGHT ATTITUDE

Most soldiers, after 3 to 4 weeks of cold-weather training, will be able to cope with the environment. Some soldiers may not be as skilled as others, but most will succeed.

Set an example for other soldiers. Winter conditions can be frightening, trying to meet challenges you have never met before. Any mistake could make you a casualty. Aggressive leadership actions that overcome the challenges of the environment are essential to mission accomplishment.

There are two enemies to overcome in northern operations: the enemy soldier and the cold. The first step in defeating either enemy is having the right attitude. The cold can defeat you psychologically if you are not aware of the symptoms of a poorly prepared unit. The following are suggestions on how to defeat the cold:

- If you become depressed or moody, or do not want to talk, force yourself to get involved with others. Do not leave out the weaker soldier.
- If you find it hard to remember things you have been taught, review orders and drills.
- Remind soldiers that their jobs are to close with, engage, and gut the enemy, and that weapons and equipment must be kept in fighting order. Operations in a northern environment can easily become a camping trip with serious consequences. You must avoid this common ambush.
- Cold weather may slow down the following of an order, but it is not an excuse for not doing it.
- Cold weather may make tasks harder and take longer to do, but it will not make them impossible. With knowledge, proper equipment, and proper training you can overcome the cold and be successful in combat.

CHAPTER 2

Personal Clothing and Equipment

"The freezing German troops were reduced to removing clothes from enemy corpses, improvising strawboots, and taking other emergency measures."

Russia, 1942

2-1. INDIVIDUAL CLOTHING

Soldiers should understand the design principles of the military cold-weather clothing system. These principles are: insulate, layer, and ventilate. There are four ways to make these principles work for you: keep it Clean, avoid Overheating, wear it Loose, and keep it Dry. By remembering the "buzz word" COLD, you can quickly check your clothing.

INSULATE

Insulation material reduces the amount of heat loss. By regulating the amount of insulation, an individual can regulate the amount of heat lost or retained.

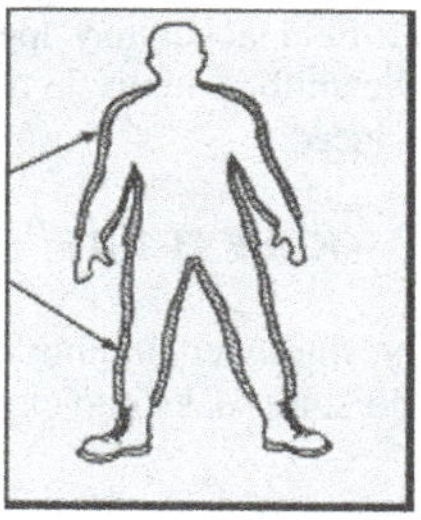

Figure 2-1. Insulation.

LAYER

Several layers of clothing provide more insulation and movement than one heavy garment, even if the heavy garment is as thick as the combined layers. Air pockets trapped between the layers provide insulation and warmth.

VENTILATE

Ventilation helps maintain a comfortable body temperature by allowing excess heat and moisture to escape.

Figure 2-2. C-0-L-D.

KEEP CLOTHING CLEAN

Dirty clothes are cold clothes. Dirt and grease clog the air spaces in clothing and reduce insulation.

AVOID OVERHEATING

Select comfortable clothing and do not overdress. In cold weather, it is better to be a little cool rather than risk being overheated.

WEAR IT LOOSE

All items of the cold-weather uniform are large enough to allow wearing of the appropriate number of layers. Your field jacket may appear too large without all of the layers designed to fit under it. If the uniform items do not fit loosely, the insulation that keeps you warm will be greatly reduced.

KEEP IT DRY

It is important to keep clothing dry, since wet clothing conducts heat away from the body more quickly than dry clothing. Moisture soaks into clothing from two directions: the inside and the outside.

From the outside you must guard against melting snow; from the inside you must guard against sweating. Brush snow and frost from clothing before entering any heated shelter or vehicle, and avoid overheating.

2-2. COLD-WET VERSUS COLD-DRY

As mentioned in Chapter 1, temperatures that are above 14°F (-10° C) are considered cold-wet conditions; temperatures consistently below 14°F will cause cold-dry conditions. Cold-wet conditions usually involve a freeze/thaw cycle, often accompanied by rain or wet snow. This type of weather is hard to dress for, since each soldier must battle the elements to keep his clothing dry. During cold conditions, the ground is usually frozen and the snow is dry. Athough the temperatures during the cold-dry conditions may be lower than during cold-wet conditions, it is easier to stay warm.

COLD-WET		COLD-DRY
	OR	
DOWN TO 14°F		BELOW 14°F

Figure 2-3. Cold-wet versus cold-dry.

2-3. THE COLD-WEATHER UNIFORM

There are many layers of clothing that must be worn in a cold weather environment. The first layer of clothing is the T-shirt and shorts. The second layer is a set of thermal underwear, which consists of an undershirt and underpants. You should wear only the clothing needed to keep comfortable and warm.

The scissors-type suspenders must be the last thing next to the outer layer, or else the layers cannot be tucked inside the trousers. The underpants and all succeeding layers of trousers are supported from the metal hooks on the suspenders. The suspenders allow the underpants and trousers to be worn loosely at the waist so that circulation and ventilation are not restricted.

Wear the wool shirt outside the trousers, which provides another method of ventilation. You can increase the waist size of this shirt by removing the stitching in each dart at the sides of the shirt.

Wear the field trousers as an outer garment in the cold-wet environment. In the cold-dry environment, wear the nylon liners inside the field trousers during periods of inactivity. On the inside of each cargo pocket on the trousers is a tape that runs through a hole on the side of the pocket. The tape ties around the thighs and keeps the cargo in the pocket from rubbing and irritating the legs. For warmth and to prevent entry of snow, tie the drawstrings around your ankles outside the boots into the boot lacing, and tuck the bowknots under the trousers.

The field jacket is worn with the liner in both cold-wet and cold-dry areas. Wear the coat and liner directly over the wool shirt. Remove the liner in mild weather or when activity increases.

In a cold-dry environment the parka is worn with a liner. The waist is adjustable for ventilation and warmth with a drawcord, and the skirt has a "split-tail" that can be tied tightly around each leg for warmth. During a period of severe cold, when activity is less, wear the parka with liner over your field jacket with liner. Remember: do not overdress or you will sweat.

Do not dry nylon on drying lines too close to the heat; too much heat will melt the nylon, destroying its insulating value.

Keep the head covered to stay warm; a warm head even warms the feet. Wear the cold-weather cap or balaclava under your helmet or under the hood to your field jacket. When wearing the cold-weather cap as an outer headpiece, the lower flap may be folded up over the top and fastened with hook-pile tape straps in front. In a cold-dry environment, wear the winter hood over the cap or the helmet. Adjust the hood by using the drawstrings on either side of the hood and the elastic strip at the back. To provide minimum protection for the face from the wind, bend the wire in the fur cuff of the hood so that you have a small opening for your face. This allows visibility but less protection.

The feet are hard to keep warm and dry, so protect them. The insulated boots (black for cold-wet conditions and white for cold-dry conditions) are warm and waterproof. The white insulated boot is the best cold-weather footwear today. When you are active, your feet will sweat and, because of the waterproof rubber lining, the moisture will remain in the socks or in the bottom of the boots. The feet, even if damp, will stay warm because the heat given off by your body will remain inside the boots. If the boots are worn for long periods, the feet become white and wrinkled when you remove the boots. This is no cause for alarm. Drying, massaging with foot powder, warming and putting on dry socks will correct this. If you are inactive for long periods, your feet may become cold. Exercises such as knee bends, stamping your feet, running in place, or wiggling your toes will make them warm. Elevate the feet whenever possible to aid blood circulation. Don't wear socks that are too tight, and wear only one pair of cushion-sole socks at a time. Change your socks at least twice a day. Most important, stay active.

Care of your boots is important. Wash boots inside and out with soapy water at least once per month or as required by unit SOP you want your feet to keep warm, you must keep the insulation dry. Inspect your shoes often and repair any holes or punctures immediately. Temporary repairs can be made by patching the hole with any kind of tape; even chewing gum can be used. Seal the hole quickly to prevent the insulation from becoming wet. If one of your boots feels heavier than the other, the insulation is probably wet and they should be turned in for new ones.

Insulated boots must be tested annually to ensure they are waterproof and have less than 0.01 percent moisture in the insulation. The air valve on the outside of the boot is used to equalize pressure when worn at high altitudes and adjusted again upon return to low altitudes. To equalize the pressure in the boot, open the valve for a few seconds and immediately close it. The valve should be kept closed at all other times to prevent moisture from entering. Never open the valve unless the boot pressure around your foot causes discomfort.

You can wear overshoes over the leather boots on wet or muddy terrain. The overshoes are not as warm and dry as the insulated boots. Make sure your socks are dry and that your boots are loosely laced so as not to restrict circulation.

2-4. EXTENDED COLD-WEATHER CLOTHING SYSTEM (ECWCS) (TEST)

The ECWCS is issued as a one-for-one replacement for the current standard cold-wet clothing system. System components use the most recent materials and technology. The result is a reduction in weight and bulk from the current cold-wet clothing system without loss of protection from the cold. The ECWCS uses the layer approach that draws moisture away from the body and should be worn as discussed below.

FIRST LAYER

This consists of polypropylene undershirt and shorts. They are worn next to the skin. They are designed to draw moisture away from the skin, thereby reducing conductive heat loss. You must make sure that no other item is placed between these items and your skin.

SECOND LAYER

The second layer contains a synthetic fiber pile (SFP) shirt and polyester batting trouser liners. The SFP shirt and trouser liners provide insulation. The shirt is designed to dry quickly if it becomes wet and to provide a high degree of insulation. It should not be worn as an exterior garment in wet or windy conditions, since the loose weave provides little protection against wind chill. If the shirt gets wet, it will reduce the insulation value.

The trouser liners are designed to be worn during little or no activity. They should not be worn during heavy activity such as skiing or snowshoeing.

THIRD LAYER

This consists of polyester batting field jacket liner and field trousers. The jacket liner should only be used for extra insulation during periods of little activity. This liner is designed to be used as a separate jacket; it should not be attached to the parka. The trousers provide a durable wind-resistant, water-repellent layer to protect the insulating underpants and trouser liners.

FOURTH LAYER

Windproof, waterproof parka and trousers make up this layer. The parka and trousers provide windproof and waterproof shell garments to protect the first three layers.

The parka should be the outer garment, which is always worn when doing any activity outside a shelter. It is waterproof, yet the fabric is breathable so that body moisture can escape. The hood is designed to give full protection from wind and snow and does not prevent your seeing the area around you. It has underarm zippers to allow air to circulate during heavy activity. The trousers are made of the same material as the parka and are very warm. They should be worn only when necessary to prevent the field trousers from becoming wet.

FIFTH LAYER

This consists of standard camouflage overwhites.

2-5. TIPS ON WEAR AND MAINTENANCE OF CLOTHING

The previous discussion on the layers of cold-wet and cold-dry uniforms describes how each system functions. In some units, the cold-dry field trouser liners are often substituted for the wool field trousers of the cold-wet system and function well in a cold-wet environment. Your commander and unit standing operating procedure (SOP) will determine what uniform and components will be worn and carried in the field.

DRYING WET CLOTHING

Hang items separately from drying lines inside the tent. Do not place anything too close to a stove or over steaming pots; steam will cause further dampness.

DRYING DAMP CLOTHING

Damp items can be hung on the rucksack during movement. Pin damp socks to load-carrying equipment or shirt during movement or just inside the last layer of clothing. Do not place damp clothing in sleeping bags. This will only make sleeping bags wet, and they are harder to dry than clothing.

Small items can be placed between the sleeping bag and the sleeping mat under your legs. When washing clothes without soap, dry-rub and air clothing and sleeping bags.

Never warm your feet next to an open fire. The boots will melt and burn because of their insulation before the feet become warm.

Try never to wear VB boots in a sleeping bag. Sometimes the tactical situation is such that boots must be worn while sleeping. If so, the feet must be dried and massaged with foot powder, and dry socks must be put on. Wearing the VB boot while sleeping does not let your feet dry out. After VB boots have dried, place them underneath your sleeping bag on top of your sleep pad when temperatures are below 0°F.

2-6. COLD-WEATHER EQUIPMENT

The sleeping equipment consists of a sleeping bag, an insulated sleeping pad, and a waterproof bag. There are two types of sleeping bags:

(1) Sleeping bag type I, intermediate cold, temperature range +45°F to +10°F, consists of a mummy-shaped bag with hood, insulated sleeping pad, and a waterproof bag.

(2) Sleeping bag type II, extreme cold, temperature range of +10°F to -50°F, is also available.

Whatever clothing is necessary must be worn with both bags to stay warm and comfortable at the lower temperature ranges. As a minimum, long underwear, sleep shirt, dry socks, and head protection should be worn while sleeping.

The insulated sleeping pad has replaced the pneumatic mattress. Use the insulated sleeping pad under the sleeping bag for warmth and comfort and to keep your sleeping bag dry.

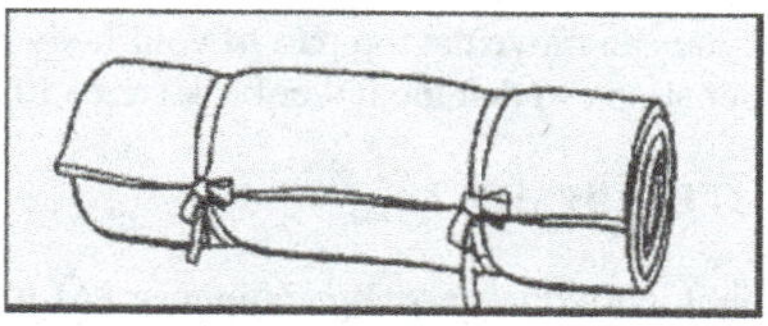

Figure 2-4. Sleeping pad.

The bag must be kept dry if it is to keep you warm. Sleeping bags become wet from outside moisture, from sweating on the inside of the bag, and from water escaping from the mouth or nose into the sleeping bag.

To keep outside moisture from wetting the bag, place protective insulation, pine or spruce boughs, a poncho, or an insulated sleeping pad under the bag. Avoid sweating by wearing the least amount of clothing to keep warm inside the bag, and by using the proper sleeping bag for the temperature conditions. Do not breathe into the bag or moisture will collect and wet the bag or form ice crystals. If your face gets cold, cover it with a towel, muffler, balaclava, or sleeping hood. The sleeping shirt with hood is specifically designed for use with the types I and II sleeping bags.

When it is not in use or being carried, open the sleeping bag wide so that fresh air can get into it. Turn it inside out so that the sun and wind can dry it. Always shake the bag out and fluff it up before and after use to provide for more trapped air between the layers of the wind-resistant/water-resistant outer shell.

Always carry your sleeping bag in the waterproof carrying bag. This protects the bag and helps keep it dry. Ensure that the top bag is securely tied to prevent snow or other moisture from getting inside and wetting the sleeping bag. The following suggestions will help you keep warm and comfortable in the sleeping bag:

- Eat just before getting into the bag. This gives you energy and helps you to stay warm.
- Always relieve yourself just before going to bed. The act of getting out of a warm sleeping bag to go outside can be a chilling experience.

PACKING YOUR RUCKSACK

When you load your rucksack for skiing or climbing, pack heavy items at the bottom and next to the frame. This places the weight on your hips, which is necessary for good balance. Place hard or sharp objects inside the load where they will not rub on the bag and your back. When walking or snowshoeing, raise the center of gravity and take more of the load on your shoulders. Articles that are often needed are placed in the outside pockets for easy reach. Keep maps and other flat objects in the flap pocket.

ADJUSTMENT

Adjust the straps so that you can move the top part of your body and swing your arms freely. Adjust the shoulder straps so that the lower back straps fit just above the belt line.

2-7. LOAD-CARRYING EQUIPMENT

The all-purpose lightweight individual carrying equipment (ALICE) rucksack comes in sizes medium and large. Make sure you are equipped with the large ALICE in a cold environment.

In cold-weather operations, you should follow the packing list set forth in your unit SOP to organize your rucksack. The main consideration is to avoid having to unpack during halts to get frequently needed items. The following illustrates a method of organizing your rucksack:

- **Small external pockets.** These contain small high energy foods (candy, cereal bars) to be eaten while on the move.
- **Large external pockets.** These contain rations to be eaten for morning and evening meals. Take extra socks, scarves, and a spare cap. The openings behind the pockets can be used for skis or other items.
- **External attachment points.** The sleeping pad is attached to the bottom of the pack or under the top flap (wrapped inside sandbags) and other combat items as required.
- **Top flap.** It contains camouflage overwhites and pack cover.
- **Main compartment.** Place the sleeping bag in the bottom of the rucksack. Spare clothes are placed in the upper half of the compartment for easy reach.
- **Camouflage.** A white cotton cover is provided that covers the entire rucksack, excluding the shoulder straps. Use it when wearing your overwhites.
- **Care and maintenance.** Check your rucksack for damages before and after an operation. Repair damages immediately or turn in through supply channels for a replacement.
- **Emergency quick-release of the rucksack.** When quick removal of the rucksack is needed, unsnap the fastener. Pull down on the quick-release fastener on the left shoulder strap, shift the rucksack toward the right shoulder, and let it drop to the ground.

- **Cargo support shelf.** The rucksack can be converted to a cargo carrier by removing the combat pack and replacing it with the cargo support shelf. You can carry square, rectangular, or box-type items such as radios, water cans, and fuel cans.

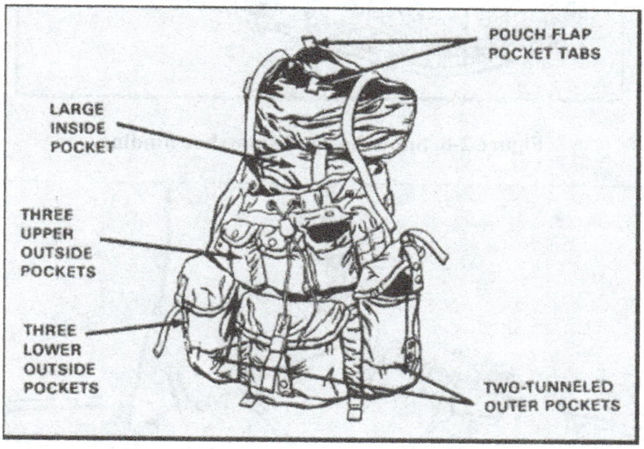

Figure 2-5. Load-carrying equipment.

2-8. OVER-THE-SNOW MOVEMENT EQUIPMENT

Before you are instructed to use snowshoes or skis, your chain command will first determine if movement by foot is possible. In shallow snow (8 inches or less) or heavily crusted snow, movement on foot for inexperienced skiers provides better maneuverability with the least effort.

SNOWSHOES

Snowshoes consist of a white magnesium frame laced with nylon-coated steel cable. They are about 48 inches by 12 inches weigh 3.3 pounds with bindings. When properly adjusted, the toe of the boot will be about halfway across the opening in the snowshoe; the toe strap should be firm, but not tight enough to restrict circulation; and the heel straps should permit some side-to-side movement. When used correctly, snowshoes provide better mobility than walking. In most instances, skis are preferred over snowshoes. When using snowshoes, you cannot back up or swim; be careful when moving over thin ice. The traction needed to ascend or descend with snowshoes can be increased by waxing. The use of ski poles with snowshoes aids movement and reduces fatigue.

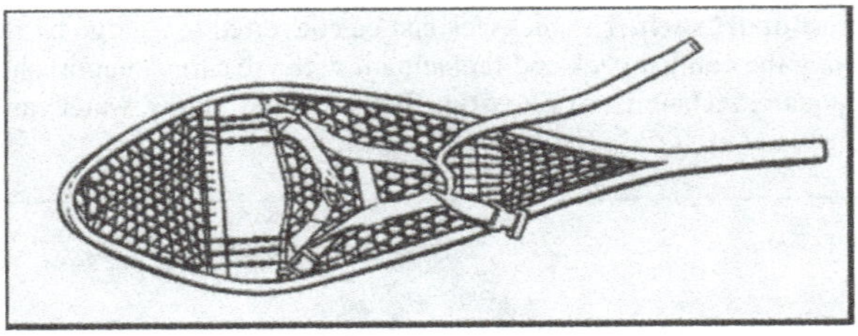

Figure 2-6. Snowshoe and snowshoe binding.

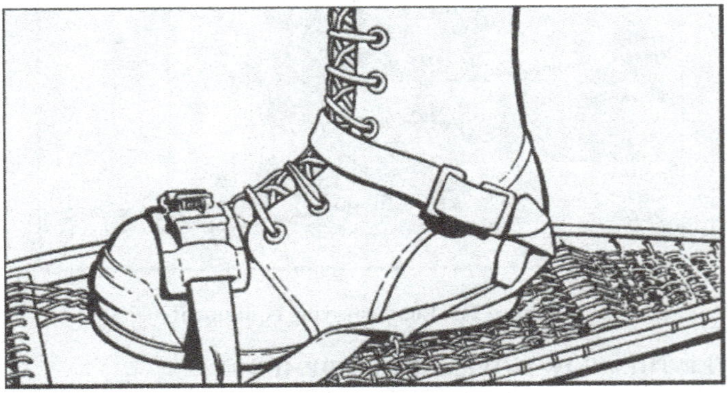

Figure 2-7. Snowshoe binding attached to VB boot.

SKIS

The standard military skis are about 208 centimeters (82 inches) in length, and have a hole in the tip for towing, a groove in the heel for attaching mohair climbers, and metal edges.

SKI POLES

Ski poles can be very useful in steep terrain. The military ski pole consists of a steel shaft with rubber handgrips and baskets, and a leather wrist strap. The poles come in three lengths: 130 centimeters (51 inches), 137 centimeters (54 inches), and 147 centimeters (58 inches). When fitting the poles, you should stand on the floor and place the pole at your side; it should come to the point of your shoulder (about 1 inch below the top of the shoulder).

MOHAIR SKI CLIMBERS

The mohair climbers are attached to the bottom of the skis for added grip when climbing steep slopes. The climbers consist of a cotton-backed mohair pile and are attached to the skis by five web straps.

2-9. MISCELLANEOUS EQUIPMENT

SUNGLASSES

Polarized sunglasses should always be worn on bright days when the ground is covered by snow. Failure to wear the sunglasses, even on overcast days when the need seems less, may result in blindness.

PROTECTIVE MASK

There is an M4 cold-weather winterization kit available for your protective mask. Make sure you install this winterization kit for operations in the northern environment.

CANTEENS

In cold weather the plastic canteen will freeze very quickly if carried in the standard fabric carrier. Since the canteen cannot be placed directly on heat, there is no quick method for thawing. Whenever possible, the canteen should be carried in one of the interior uniform pockets, or wrapped in clothing and placed in the rucksack to ride against your back. In extreme cold, do not fill the canteen over two-thirds full; this allows for expansion if the water freezes.

The insulated canteen is a one-quart capacity stainless steel vacuum bottle. It is provided with a cotton duck cover and a stainless steel cup. A new arctic canteen will keep water from freezing at -40°F for 6 hours. The following are some tips in using the arctic canteen:

- Fill the canteen using your canteen cup.
- Fill to within half an inch of the neck.
- Do not drop or dent.
- Put it in your sleeping bag at night.
- Carry the canteen in an M60 bandoleer near your body.

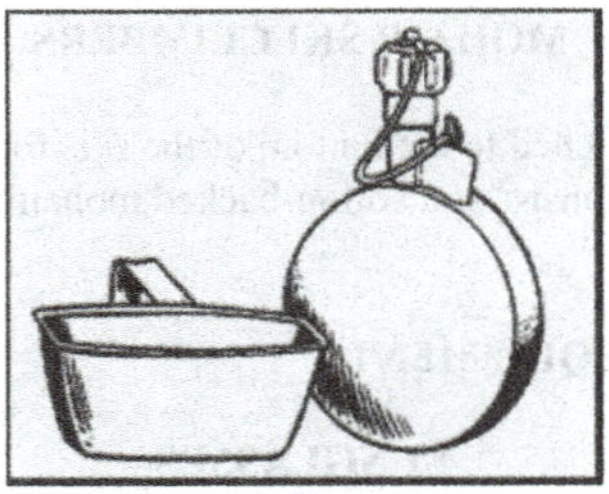

Figure 2-8. Insulated canteen with cup.

CHAPTER 3

Tents and Heating Equipment

"A well-sheltered defender enjoyed a marked advantage over his exposed attacker..."

Ironside, 1918

3-1. GENERAL

To conduct successful military operations in cold weather and maintain a high level of combat efficiency, you must have heated shelters. Tents and stoves are a vital part of cold-weather equipment.

3-2. TENT GROUP EQUIPMENT

The following is a list of the tent or ahkio group for an infantry squad:

1 Ten-man tent with liner
1 Yukon stove
1 Metal or plywood stove base
2 Five-gallon gasoline cans (one with white gasoline per platoon)
1 Five-gallon water can (filled)
2 Cases, meals, ready-to-eat
1 Climbing rope (120 feet)
1 Box of candles
2 Squad cook sets
2 One-burner squad stoves
2 Axes
2 Saws
1 Hatchet with sheath
1 D-handle shovel
1 Sledgehammer
1 Pick
1 Tripod for gas can (set)
3 Pairs of snowshoe bindings (spare)
1 Lantern (extra mantles)
1 Box of soda bicarbonate (fire extinguisher)
1 Roll of soft wire
5 Plastic bags
3 Rolls of toilet paper

3-3. AHKIO

The ahkio is the squad's method of transporting the tent group equipment in the winter. It is a 200-pound capacity fiberglass sled that can be used for transporting tent group equipment, for carrying weapons and ammunition, as a firing platform, and for the evacuation of casualties.

3-4. PACKING

When packing the ahkio, the load must be evenly distributed. Place heavy equipment on the bottom and slightly to the rear of center. Loading lighter equipment toward the top will prevent the ahkio from becoming top heavy. Tools such as shovels, axes, and saws are packed at the sides for easy access in trailbreaking; however, caution must be taken to ensure that these items do not stick out. The canvas cover is folded over the load. To keep the load from shifting and to prevent snow from getting under the cover, lash the load tightly by crisscrossing the lashing rope from the lashing rings on one side of the sled to the rings on the other side. An additional ahkio, issued on a basis of one per two 10-man tents, should be used to carry camouflage nets and support systems. In very wet or dry snow, the movement of an ahkio is easier when wax is applied to the running surface.

3-5. TEN-MAN ARCTIC TENT

The 10-man arctic tent with liner is a six-sided pryamidal tent supported by a telescopic pole. The tent is designed to accommodate 10 men and their clothing and equipment. It will hold additional personnel by leaving packs and equipment outside the tent, or by lowering the center pole, which allows the sidewalls to spread out to cover more ground surface. The tent has two doors, which permits joining of tents for access from one tent to another. A snow cloth is attached to the bottom of the sidewalls for sealing the tent to the ground. The snow cloth should always be to the outside of the tent with pine or spruce boughs underneath, or it may freeze to the ground. This allows easy escape in case of an emergency if the doors are blocked. The tent is ventilated by four built-in ventilators on opposite sides and near the top of the tent.

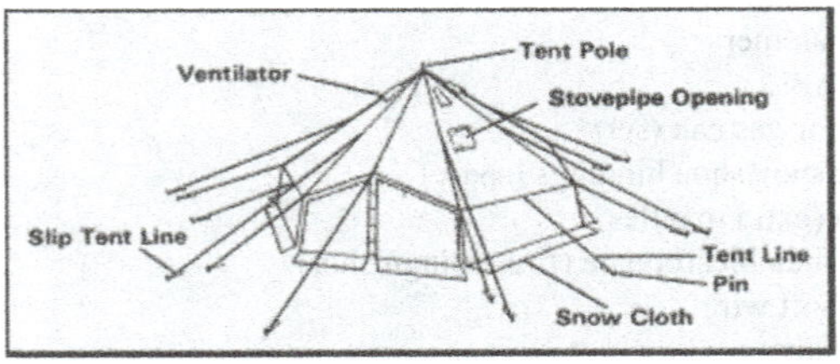

Figure 3-1. Ten-man arctic tent.

3-6. BUILDING ARCTIC TENTS

The shelter is easily built by four trained soldiers. Consider the following when pitching or striking tents in snow or on frozen ground.

Snow should be cleared to the ground surface to obtain a lower silhouette and gain advantage of ground temperatures, which are generally warmer than upper air temperatures. Enough pine or spruce boughs should be placed on the tent and floor, and under the snow covers, for insulation and comfort. If an insulating material is not used, the heat of the stove will thaw the ground insulation in a sea of mud. The snow covers will freeze to the ground, preventing escape from the tent in case of an emergency. When it is impractical to remove the snow to the ground level, an adequate tent site can be prepared by packing the snow with skis or snowshoes until a firm base is provided for pitching. The center pole is placed on a log or other suitable support to keep it from sinking into the snow.

In open terrain, with a strong wind, it will be necessary to build a 1-meter-high snow wall on the windward side of the tent to protect it from the wind. The tent is pitched with the entrance angled about 45 degrees from the downwind side. When the tent is set up, the snow cloth around the bottom of the tent should be spread out so that snow or logs can be placed on it to help in anchoring the tent.

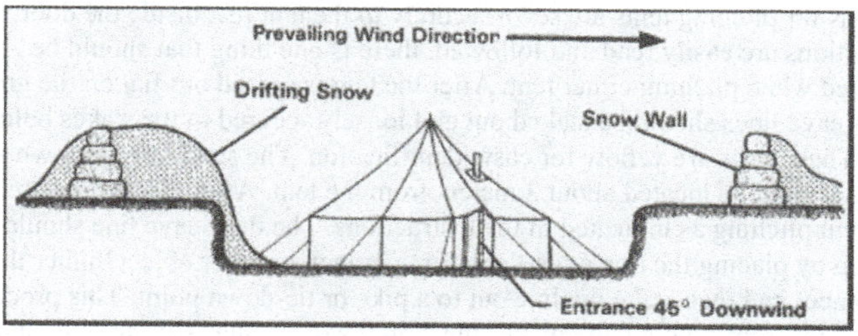

Figure 3-2. Windbreak.

Driving tent pins into frozen ground is impossible. Tents may be pitched and anchored by attaching the tent lines to trees, branches, logs, or stumps. If these are not available, holes are dug into the snow for the purpose of using "deadmen." This is done by digging a hole into the snow large enough to insert a pole or log about 1 meter long with the tent line attached. The hole is filled with snow and well packed. The packed snow freezes, securely anchoring the tent. In powder or dry snow, water would be poured into the hole.

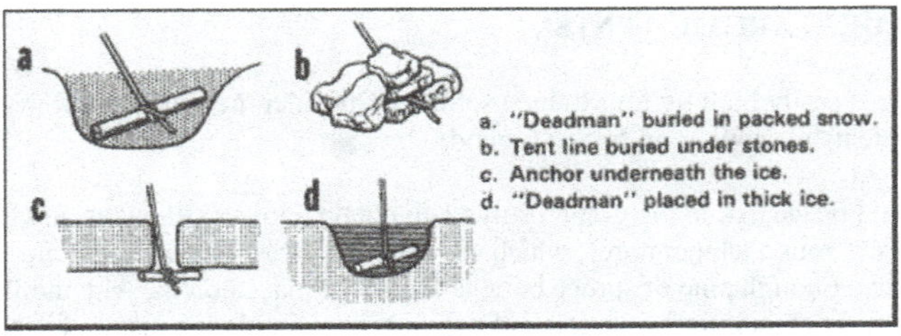

Figure 3-3. Deadmen.

The smaller tent (five-man) provides the commander more flexibility in distribution of his forces. You can never be assigned a tactical mission in the arctic without provisions for a shelter. This tent enables a smaller unit to be assigned an independent mission, such as patrol or reconnaissance mission. It is also six-sided, pyramidal, and supported by a telescopic tent pole. It is designed to accommodate four to five men and their clothing and equipment. The tent can provide shelter for a rifle squad or other similar unit when rucksacks are placed outside the tent.

3-7. INSTRUCTIONS FOR PITCHING TENTS

Instructions for pitching tents are sewn securely to the tent just inside the door. Although the instructions are easily read and followed, there is one thing that should be remembered when pitching either tent. After the tent is spread out flat on the ground, the six corner eave lines should be staked out and loosely secured to the stakes before the tent is raised. These lines are yellow for easy identification. The stakes or tie-down points for these lines should be located about 2 meters from the tent. After this is done, proceed with the tent pitching as indicated in the instructions. The door eave line should be propped up by placing the line over a pole (tree branch or other object higher than the door entrance) and then stake the line out to a pike or tie-down point. This procedure keeps the door from sagging and makes the zipper work better. Drying lines should be attached to the liner inside the tent to allow drying of clothes.

In winter the tent will normally be covered with ice and snow, which must be removed before the tent is struck. This can be done by shaking the tent or by beating it with a mitten or stick. If the snow cloth is frozen to the ground, the snow and ice around it must be removed by chopping or shoveling. This must be done carefully to avoid damage to the snow cloth. One way to do this is to ease the shovel between the cloth and the ground and gently pry the cloth away from the ice.

When the bottom of the tent is properly sealed and the door is zipped, moisture will form on the inside of the tent and collect on clothing and equipment. Carbon monoxide and fumes from the stove can accumulate to a dangerous degree. To avoid this, ensure that the ventilators near the peak of each tent are open.

The roof of the shelter must be steep enough to shed snowfall. With both the 5-man and 10-man tent, the area that will become the floor of the tent is dug, packed, and insulated with pine, spruce boughs, or similar material. Snow walls should then be built to protect against wind.

3-8. YUKON STOVE

The Yukon stove is used to heat the 10-man, 5-man, and GP small tent. In addition to providing heat, the top surface of the stove and the area beneath the stove may be used to heat rations and water. The stoves use standard MOGAS but may also be operated with diesel fuel or JP-4 without modification. NEVER MIX FUELS. When solid fuels (wood, coal, etc.) are used, the stove must be modified by removing the burner unit, closing the burner opening, and turning over the wire grate so there is space below for draft and ashes.

WARNING

DO NOT USE A SPARK ARRESTER ON THE M1950 YUKON STOVE.

A flue cap, issued as a component part of the Yukon stove, shields the stovepipe top from the wind and prevents the wind from blowing smoke or gases down the pipe into the stove and tent. The flue cap is equipped with three 4 1/2-meter lines that are used to steady the stovepipe and anchor it in high winds. These guy lines must be anchored to the tent or tent ropes, not to the ground or nearby trees. The stovepipe hole cover must be rolled up and tied off to prevent nylon ties from touching the stovepipe.

A firm base for the stove must be provided to keep it from melting down into the snow. The base should be 35 centimeters by 70 centimeters and be made of lightweight metal or salvaged plywood. If plywood is used, it should be covered with metal or aluminum foil. If those materials are not available, a base can be made out of small logs about 10 centimeters in diameter.

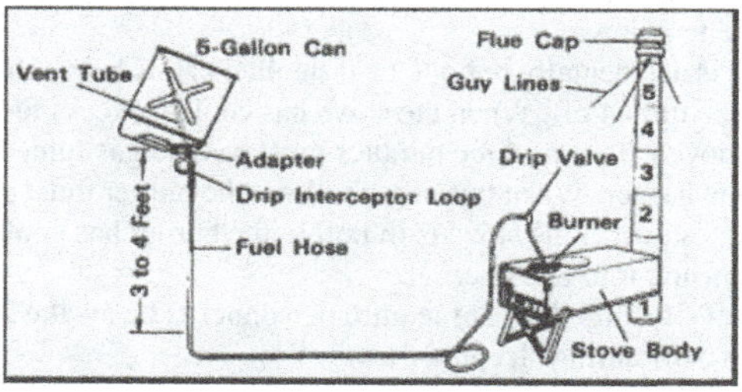

Figure 3-4. Yukon stove.

A simple stand for holding the fuel can is built by obtaining three poles 2 meters long. The poles are tied about two thirds of the way up and then spread out to form a tripod. The fuel can should be at least 1 meter higher than the stove. If the fuel can is wobbly or if there is a wind, the can should be tied to this tripod. Make sure the can is tilted so that the air is trapped in the uppermost corner.

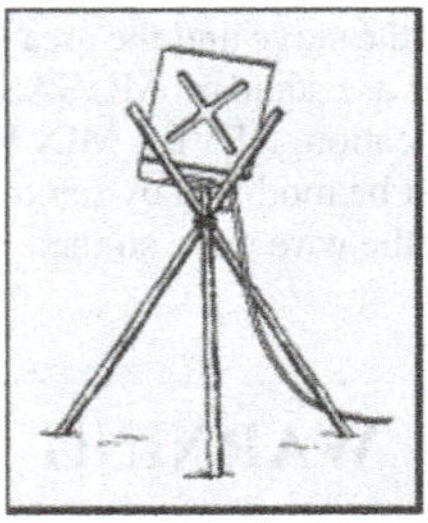

Figure 3-5. Tripod for gas can.

3-9. PRECAUTIONS

The following precautions must be observed when the Yukon stove is being used:

- All stovepipe connections must be tight, and the tent stovepipe shield properly adjusted. The stove must be level to ensure that the burner assembly will spread the flame evenly throughout the stove.
- Ensure that the fuel line is made of natural rubber and suitable for extreme-cold operation. It must be protected so it cannot be accidentally pulled loose and must have a drip intercept loop to prevent fuel from running down the line and into the tent.
- Rate of fuel flow must be checked regularly. It will change with the fuel level and will need adjustment. All fuel supplies must be kept outside the tent.
- The stove should never be left unattended when it is burning. Allowing the stove to become too hot may cause the stove body to become overheated and warped and burn the tent down.
- If the flame is accidentally put out, or if the fuel can is being changed, the drip valve must be turned off. When the stove has cooled, excess fuel inside the stove must be removed. Two or three minutes must pass so gas fumes escape before relighting the burner. When burning gasoline, the burner must be cool before relighting the stove. If the stove is lit before the burner has cooled, the fuel will vaporize, causing it to explode.
- If the tactical situation does not require personnel to be awake 24 hours a day, turn the stove off during sleeping time.

The following are precautions when **burning wood or coal**:

- Fuel should be fed in small amounts until there is a hot bed of coals.
- The stove should not become overheated.
- Gasoline should not be poured on the wood or coal inside the stove to hurry starting the fire.
- Ashes should be removed regularly and not allowed to accumulate higher than the bottom of the grate.
- NEVER MIX FUELS.

3-10. SQUAD STOVE, M1950

The one-burner squad stove is a cooking and heating unit for two to five men. It is used when operating in an isolated or forward area when separated from the ahkio group. The stove is small, compact, and lightweight, and it will operate on either MOGAS or white gasoline. The initial pressure for its operation is generated by a few strokes on the hand pump. Thereafter, the pressure is generated by heat from the burner evaporating the fuel in the stove tank.

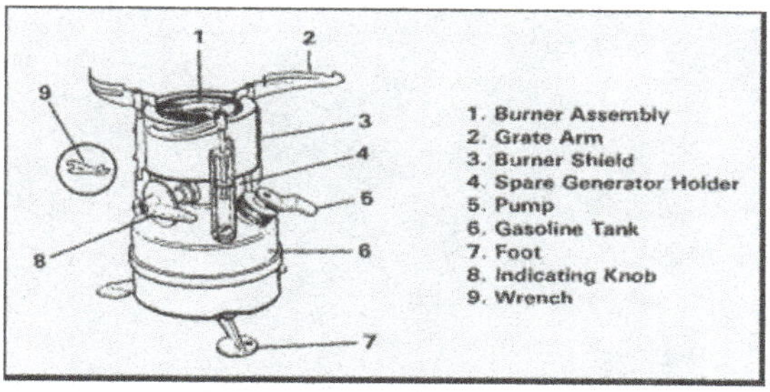

Figure 3-6. Gasoline cooking stove, M1950.

To operate the squad stove, do the following:

- Snap out the three feet and set stove on a level surface.
- Raise the grate arms and turn them outward so that the inner tabs can be engaged in the slots in the burner shield.
- Close the valve by turning the indicating knob clockwise to the OFF position.
- Unscrew the pump caps and remove the pump assembly.
- Fill the gasoline tank three-fourths full.
- Wipe off any gasoline that may have been spilled during filling and replace the pump cap.
- Pump 10 strokes to build up initial pressure.
- Turn the indicating knob to the ON position for 3 seconds. This allows the gasoline to rise through the generator to the burner and preheated cup.

- Turn the indicating knob to the OFF position to prevent excess fuel from accumulating in the burner assembly.
- Light the wick and allow the flame to burn for about 3 minutes or until the flame turns blue.
- Turn the indicating knob to the LIGHT position. Have a second match ready to relight the burner immediately if the flame goes out.
- When the flame in the burner turns blue, turn the indicating knob to the ON position.
- Pump an additional 10 strokes to maintain pressure within the tank. The stove is now ready for use.
- If the flame burns unevenly or dies during operation, turn the valve to the OFF and ON position several times, relight quickly, and pump several additional strokes.
- To turn the stove off, turn the indicating knob to the OFF position and allow the flame to go out.

CHAPTER 4

Small Arms and Ammunition

"Rifles and machine guns and even the breechblocks of artillery became absolutely rigid ... and steel parts cracked."

General Rendalic, Eastern Front, 1942

4-1. GENERAL

Cold weather will greatly affect the operation of your weapon. You must be aware of these effects to ensure employment of your weapon under poor weather conditions.

4-2. CARE, CLEANING, AND MAINTENANCE

Your weapon will work properly under extreme cold conditions if you take care of it. Normal temperate zone lubricants thicken in cold weather, and stoppages or sluggish actions of your weapon will result. During the winter, disassemble your weapon completely, clean it thoroughly, and remove all lubricants. Oil the entire weapon with a light coating of cleaner lubricant preservative (CLP).

> **CAUTION**
>
> **DO NOT USE CLP ON A MORTAR OR ANY COMPONENT PART OF A MORTAR.**

Figure 4-1. Weapons will 'sweat" when taken in to a heated shelter.

A main problem is to keep snow and ice, which may disable your weapon, out of the working parts, sights, and barrel. When the weapon is not in use, keep it out of the snow either in an improvised weapon stand or in a tripod arrangement. If this is not possible, stand it up with the butt end in the snow.

Condensation ("sweating") will form on the weapon when you take it from the cold into a heated shelter. The condensation will continue for about 1 hour, after which you should clean the weapon. If you don't, the condensation will freeze when you take the weapon outside, and it may not fire. The weapon must be kept as near the floor as possible to lessen condensation. If parts of the weapon do become frozen, warm them and move the working parts slowly until they are unfrozen. If you can't warm your weapon, remove all ice and snow and move the parts gently until action is restored. The use of CLP will prevent malfunctions by eliminating condensation.

When firing, do not let the hot parts of your weapon come in contact with the snow. The snow will melt and, as the weapon cools, will form ice. When changing barrels, don't lay them on the snow.

To prevent bipods and tripods from sinking into the snow, improvise for added flotation. This is done by placing them on snowshoes, ski poles, branches, ahkios, etc. Baseplates should be on solid ground or sandbags to prevent sinking into the snow and damage to the weapon.

Figure 4-2. Bipods and tripods will sink in the snow.

AMMUNITION

Small arms ammunition should be kept at the same temperature as your weapon. Ammunition clips and magazines must be cleaned of all oil and preservative and checked often. All ice, snow, and condensation must be removed. Ammunition boxes must be kept closed in order to keep out snow and ice.

Stored ammunition should be kept in packed boxes, raised off the surface by placing them on boughs, cardboard, or other similar material, and covered to protect them from the snow. Ammunition should be well marked for easy location and identification in case it becomes covered with snow.

Ammunition resupply may become restricted. Everyone must be aware of the necessity for ammunition economy and fire discipline. Loaded clips, magazines, or single rounds dropped in the snow become quickly lost; therefore, careful handling of ammunition is essential. Never waste your ammunition. Make every round count.

When contact with the enemy is imminent, troop compartments and aircraft should be kept as close to freezing as possible to avoid condensation. However, many munitions have limitations at lower temperatures, which must be recognized in training.

CHAPTER 5

Rations and Diet

> ***"I ate my boots and a pair of pants ... tea and seal-skin gloves for dinner."***
>
> ***Greely, Arctic Expedition, 1883***

5-1. GENERAL

Most of what we eat and drink is used in maintaining our body heat, while only a small portion is used in producing energy for physical work. You must ensure adequate caloric intake in cold-weather operations. About 4,000 calories per day are necessary for personnel performing physically demanding work in the cold. Efficiency may drop rapidly if this level is not maintained.

The body loses liquid at a very fast rate in arctic conditions, regardless of how carefully you adjust and ventilate your clothing. The exertion of movement on foot, preparation of bivouac sites and defenses in the snow, etc., take a toll in sweat and loss of moisture in the breath. These liquids must be regularly replaced, preferably by hot drinks, which provide extra calories if they contain sugar.

5-2. RATIONS

Rations provide the needed calories to live and fight effectively. When eaten in their entirety, rations contain the right amount of carbohydrates, fats, protein, and vitamins. The proper intake of these essential items depends on the entire ration being eaten in properly spaced meals. The lack of concern regarding eating caused by the cold, combined with the difficulties and inconvenience of cooking, may tempt soldiers to miss meals. The principles of sound leadership and discipline in cold weather require that meals be prepared and that the entire ration be eaten, warmed when possible. Save snacks (cookies, chocolate bars, etc.) for between meals and when on the march.

5-3. LIQUIDS

Because there is a large amount of fluid loss in the cold, you should drink a minimum of 3.5 quarts of water a day when heavy physical activity is involved. Water is usually available either from streams or lakes or by melting snow or ice. A limiting factor may be fuel needed to melt the snow or ice. To save time and fuel, water should be obtained from running streams or a lake. The milky water of glacial streams must be allowed to stand until the sediment settles. When a hole is cut into the ice to get water, it should be covered by a snow block or something similar to slow refreezing.

Ensure that as much of the daily liquid intake as possible is hot drinks, such as soups and cocoa. Your local command may request a special food allowance of warming and cooling beverages. Main meals should begin with soup, and between-meal snacks should include a hot drink.

When running water is not available, ice or snow must be melted. Ice produces more water in less time than snow. When melting snow, first put a small amount into the cooking pot; then more can be added. Continue this process until there is enough water. It must be purified by boiling rapidly for 15 minutes.

When heating water, use all available stoves, since this is a most time-consuming operation. Melting and boiling enough snow for a drink may take 30 to 40 minutes. Areas designated for ice or snow that is to be used for water must be sited far upwind of the latrine and garbage disposal sites. Before going to bed, ensure that stoves are filled and enough water has been prepared for breakfast. Do not let other soldiers eat snow or ice; doing so lowers the body's core temperature.

Never drink alcohol on operations or when exposed to low temperatures. The aftereffects of alcohol can lead to a feeling of false security. You may forget the main rules to prevent cold injury, like wearing your gloves or not going outside without shoes to relieve yourself.

CHAPTER 6

Hygiene and First Aid

"...the first frostbite cases have made their appearance. We amputated two feet and will amputate many more."

Amosoff

6-1. GENERAL

The care of your body needs special attention in cold weather. Ensuring cleanliness in the field, particularly in the cold, is important. It is easy to let yourself go for long periods without bathing, but if you wish to be part of an effective combat force, you must keep clean.

6-2. HYGIENE TIPS FOR THE SOLDIER

Face, hands, armpits, and crotch should be washed daily. If water is not available, take a "bath" using cornstarch or rub your body briskly with a dry towel. The cornstarch removes excess oil and perspiration from skin and hair.

Shave regularly -- daily if water is available or every 2 to 3 days if water is in short supply. Shaving just before sleeping will allow the face maximum recovery time before going out into the cold, reducing the possibility of superficial frostbite.

Teeth should be cleaned daily. If a toothbrush is not available, a clean piece of cloth wrapped around a finger or a chewed twig may be used as a substitute.

Underwear should be changed at least twice a week. If it is not possible to wash your underwear, crumple it, shake it, and air it for at least 2 hours.

Ensure that all eating and cooking utensils are cleaned after each use.

In the dry arctic you will find the need to oil your unprotected and exposed skin once every week (with an oil such as suntan oil).

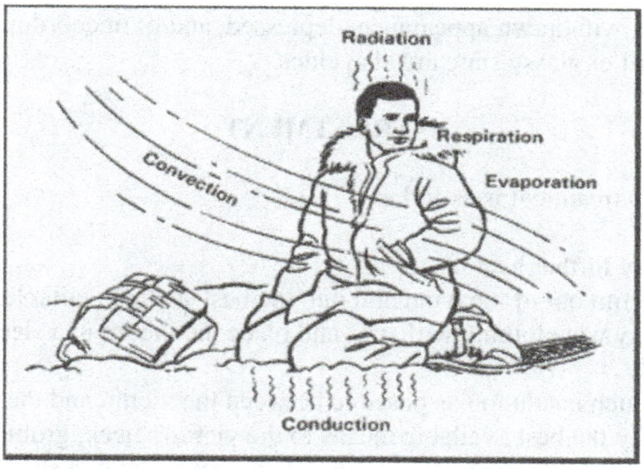

Figure 6-1. Mechanisms of heat loss from the body.

6-3. HYPOTHERMIA

Hypothermia is a lowering of the temperature of the body's inner core. This happens when the body loses heat faster than it can produce it. You must be able to recognize the symptoms of hypothermia and to treat the condition immediately.

PREVENTION

To prevent hypothermia, adhere to the following:

- Stay physically fit.
- Keep active.
- Use the uniform properly and keep it dry.
- Eat properly and often.
- Drink plenty of liquids, at least 3.5 quarts per day, when performing physically demanding work.
- Be prepared for and know how to deal with quick changes in the weather.
- Bivouac early before judgment is weakened by fatigue.

SYMPTOMS

Be aware of the following symptoms:

- Shallow breathing or absence of breathing.
- Faint or undetectable pulse.
- Victim is first cold, then stops shivering.
- Loss of feeling.
- Poor coordination; victim has trouble walking.
- Impaired speech.

- Confusion, withdrawn appearance, depressed, and/or uncoordinated. An uncaring attitude and/or glassy stare are also clues.

TREATMENT

The recommended treatment is as follows:

- Prevent any further heat loss.
- Get the victim out of the wind and into the best shelter available.
- Replace any wet clothing with dry, and place the victim in a sleeping bag, if one is available.
- Place as much insulation as possible between the victim and the ground.
- Add heat by the best available means to the victim's neck, groin, and sides of chest. This can be done by using a hot water bottle, Yukon stove, campfire, or your own body heat.
- Provide the victim with something warm and nutritious to drink if he is conscious. Calories may be added by using sugar in hot, sweet drinks.
- DO NOT massage the victim.
- DO NOT give alcohol to the victim.
- Evacuate the victim to the nearest medical treatment facility as soon as possible. This condition is life threatening until normal body temperature is restored.

6-4. FROSTBITE

Frostbite results when tissues freeze from exposure to temperatures below 32°F. The degree of injury depends upon the wind-chill factor, length of exposure time, and proper protection. Persons with a history of cold injury are prone to frostbite. There may be a tingling, stinging, an aching sensation, or a cramping pain. The skin first turns red and then becomes pale gray or waxy white. Frostbite can be classified as superficial or deep.

WIND CHILL

The combination of wind and low temperatures creates a condition known as wind chill. For example, with the wind calm and a temperature of -20°F there is little danger of wind chill. If the temperature is -20°F and there is a wind of 20 knots, the equivalent chill temperature is -75°F!

Under these conditions there is GREAT DANGER and your exposed skin can freeze within 30 seconds. You also create your own wind when you walk, run, or ski. As examples, walking creates a wind speed of about 5 mph and running 10 mph. Wind is also created by moving vehicles, aircraft propellers, and rotors. The wind-chill chart is provided to show how wind and temperature combine to create wind chill; use it. Cold injury can disable you like an enemy bullet.

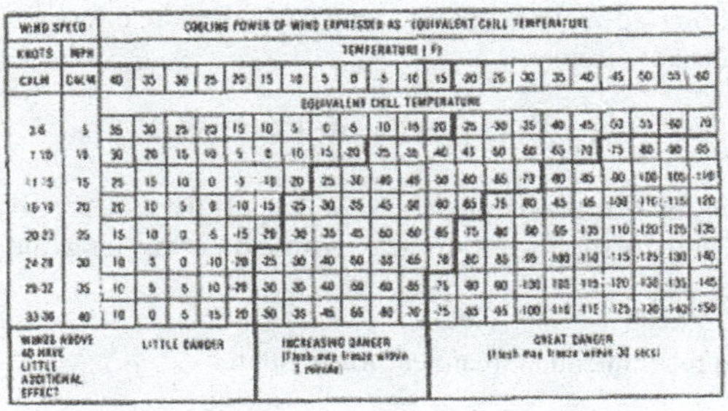

WIND SPEED		COOLING POWER OF WIND EXPRESSED AS "EQUIVALENT CHILL TEMPERATURE"																				
KNOTS	MPH	TEMPERATURE (°F)																				
CALM	CALM	40	35	30	25	20	15	10	5	0	-5	-10	-15	-20	-25	-30	-35	-40	-45	-50	-55	-60
		EQUIVALENT CHILL TEMPERATURE																				
3-6	5	35	30	25	20	15	10	5	0	-5	-10	-15	-20	-25	-30	-35	-40	-45	-50	-55	-60	-70
7-10	10	30	20	15	10	5	0	-10	-15	-20	-25	-35	-40	-45	-50	-60	-65	-70	-75	-80	-90	-95
11-15	15	25	15	10	0	-5	-10	-20	-25	-30	-40	-45	-50	-60	-65	-70	-80	-85	-90	-100	-105	-110
16-19	20	20	10	5	0	-10	-15	-25	-30	-35	-45	-50	-60	-65	-75	-80	-85	-95	-100	-110	-115	-120
20-23	25	15	10	0	-5	-15	-20	-30	-35	-45	-50	-60	-65	-75	-80	-90	-95	-105	-110	-120	-125	-135
24-28	30	10	5	0	-10	-20	-25	-30	-40	-50	-55	-65	-70	-80	-85	-95	-100	-110	-115	-125	-130	-140
29-32	35	10	5	-5	-10	-20	-30	-35	-40	-50	-60	-65	-75	-80	-90	-100	-105	-115	-120	-130	-135	-145
33-36	40	10	0	-5	-15	-20	-30	-35	-45	-55	-60	-70	-75	-85	-95	-100	-110	-115	-125	-130	-140	-150
WINDS ABOVE 40 MPH HAVE LITTLE ADDITIONAL EFFECT		LITTLE DANGER					INCREASING DANGER (Flesh may freeze within 1 minute)						GREAT DANGER (Flesh may freeze within 30 secs)									

Figure 6-2. Wind-chill chart.

PREVENTION

Frostbite is preventable if the individual soldier takes care of himself. It seldom occurs in people who are maintaining enough body heat. It is most commonly associated with an overall body heat loss resulting from poor equipment, reduced food intake, dehydration (water loss), overtiring, injury, or a combination of these factors. The feet, hands, ears, and exposed areas of the face are most easily hurt by frostbite and must receive constant care. The following are rules to help prevent frostbite:

- Do not wear tight boots and socks.
- Use the "buddy system" to check exposed areas, especially when wind is present.
- Carry extra socks and mitten liners.
- Do not stay still for long periods.
- Use caution when cold and wind are combined.
- Check feet during halts.
- Ensure that you eat well and have hot drinks as often as possible.

SYMPTOMS OF SUPERFICIAL FROSTBITE

Redness occurs, followed by powdery flaking of the skin. Affected areas of dark-skinned soldiers may appear dull and grayish. Blister formation occurs 24 to 30 hours after exposure, followed by flaking of superficial skin.

SYMPTOMS OF DEEP FROSTBITE

Common symptoms are as follows:

- Loss of feeling occurs in the affected area, leaving it pale yellow and waxy looking.
- When the frostbitten area thaws, it is painful.
- Frozen tissue may feel solid or "wooden" to the touch.
- When exposed to inside temperatures, blisters may appear in 12 to 36 hours.

- Discoloration (red-violet) appears 1 to 5 days after the injury.
- Gangrene usually results.

TREATMENT FOR FROSTBITE

Determine whether the frostbite is superficial or deep. If the exposure time was short, the frostbite will probably be superficial. If the exposure time was long, the frostbite will probably be deep.

The following is a recommended treatment for frostbite:

- Move the casualty to a warm and sheltered area.
- Do not rub the frostbitten area with snow or apply cold water soaks.
- Do not rewarm the affected area by massaging or exposing it to open fire.
- Rewarm the face, nose, or ears by placing your hands on the frozen area. Rewarm frostbitten hands by placing them under clothing and against the body.
- Close the clothing to prevent further loss of body heat.
- Rewarm the feet by removing the boots and socks. Place the bare feet under the clothing and against the abdomen of a buddy. Once the feet are warmed, put on dry socks and boots, if available. If the victim must wear the wet socks and boots, he should exercise his feet by wiggling his toes.
- Loosen tight clothing and remove jewelry.
- Improve circulation by exercise.
- Do not allow the victim to use alcohol or tobacco. Alcohol increases loss of body heat; tobacco causes narrowing of blood vessels in the arms and legs.
- Reassure the victim.
- Protect frozen tissue from further cold or trauma, and evacuate the casualty to the nearest medical treatment facility.

Deep frostbite is most common and harmful to the feet and less common to the hands and ears. When deep frostbite does occur, the victim should be moved to a sheltered area and evacuated to a medical treatment facility immediately. If possible, do not allow the victim to walk if his feet are frozen and avoid treating or thawing the affected area. Thawing of deep frostbite in the field increase pain and invites infection, greater damage, and gangrene. Do not try to thaw deep frostbite; let the physician do it.

6-5. DEHYDRATION

Dehydration can cause serious physical problems and make you more susceptible to other problems, such as frostbite and hypothermia.

PREVENTION

The following are suggestions to prevent dehydration:

- The **minimum** daily liquid requirement for personnel performing demanding physical work in the cold is 3.5 quarts per day.
- By the time you feel thirsty you are already dehydrated. Drink whenever possible, particularly during halts.
- Ensure that canteens are full before any type of movement.
- During periods when water is scarce, do not drink coffee in large quantities. It will pass through you quickly. Water should be the first choice.
- Check your urine spots in the snow. Dark yellow or brown indicates dehydration.

SYMPTOMS

Common symptoms are as follows:

- Red indicates a severe condition needing immediate medical evacuation to the nearest medical treatment facility.
- Lack of appetite.
- Dry mouth, tongue, and throat.
- Stomach cramps and/or vomiting.
- Headaches.

TREATMENT

Recommended treatment is as follows:

- Keep the victim warm.
- Give him plenty of liquids.
- Make him rest.

6-6. CARBON MONOXIDE POISONING

Whenever a stove, fire, gasoline heater, or internal combustion engine is used indoors, there is danger of carbon monoxide poisoning. Fresh air in living and working quarters is vital. Carbon monoxide is a deadly, odorless gas.

PREVENTION

The following are suggestions to prevent carbon monoxide poisoning:

- Use stoves and lanterns in well-ventilated areas and tents.
- Ensure that stoves and lanterns are functioning properly.
- Do not let personnel warm themselves by engine exhaust.
- Always have windows cracked in vehicles with a heater in use.

- Use a tent guard or shut the stove off when sleeping.

SYMPTOMS

Common symptoms are as follows:

- Headache, dizziness, confusion, yawning, weariness, nausea, and ringing in the ears.
- Bright red color on lips and skin.
- Victim may become drowsy and collapse suddenly.
- If personnel are found unconscious in an enclosed shelter, carbon monoxide poisoning should be suspected.

TREATMENT

Adhere to the following for proper treatment:

- Move the victim to open air.
- Keep the victim still and warm.
- If the victim is not breathing, administer mouth-to-mouth resuscitation.
- Administer cardiopulmonary resuscitation, if necessary.
- Immediately evacuate the victim to the nearest medical treatment facility.

6-7. SNOW BLINDNESS

Snow blindness is caused by ultraviolet and ultrablue rays of the sun being reflected from a snow-covered surface into the eyes. This condition can occur even in cloudy weather. In fact, it is more likely to occur in hazy, cloudy weather than in sunny weather.

Figure 6-3. Snow blindness.

PREVENTION

Prevention is easy; wear your issued sunglasses. If sunglasses are not available, make slitted glasses from cardboard, thin wood, tree bark, or similar material.

SYMPTOMS

Common symptoms are as follows:

- A scratchy, sandy feeling under your eyelids.
- Redness and watering of the eyes.
- A headache.

TREATMENT

Adhere to the following for proper treatment:

- Blindfold the patient using a dark cloth.
- Reassure the patient.
- Evacuate the patient to a medical treatment facility.

6-8. SUNBURN

Sunburn is a serious hazard because of the reflective qualities of the snow, especially at higher elevations.

PREVENTION

Require soldiers to use a sunscreen and lip balm.

SYMPTOMS

Redness of skin with slight swelling is one symptom of sunburn. Prolonged exposure to the sun may cause pain and blistering and, in severe cases, chills, fever, and headaches.

TREATMENT

Soothing skin creams may be helpful if the swelling is not severe.

6-9. TENT EYE

Tent eye is an inflammation of the eye caused by fumes from stoves and lanterns used in a poorly ventilated tent or shelter. It is prevented by proper ventilation of tents or shelters. The treatment is fresh air.

6-10. TRENCH FOOT/IMMERSION FOOT

Trench foot and immersion foot are injuries caused by the prolonged exposure of skin to cold or wet conditions. In extreme cases, the skin dies and amputation of the foot or leg may be necessary.

PREVENTION

The following helps prevent trench foot or immersion foot:

- Change to dry socks at least twice daily.
- Dry and massage your feet regularly with foot powder.
- Wear VB boots and change socks regularly.
- If leather boots are worn, dry them whenever possible.

SYMPTOMS

In early stages, the feet and toes are cold, numb, and still, and walking becomes challenging. The feet swell and become painful.

TREATMENT

In early stages, keep the feet dry, clean, and exposed to the air. In later stages, evacuate the victim to a medical facility.

6-11. CONSTIPATION

At very cold temperatures, personnel tend to put off the natural urge to have a bowel movement, resulting in constipation. This can be a painful casualty producer.

PREVENTION

Eat fresh or canned fruit and eat regularly. Drink plenty of liquids. Try to have a daily bowel movement at your normal time.

SYMPTOMS

Some common symptoms are stomach cramps, dizziness, and headaches.

TREATMENT

Prevention is the best treatment. If symptoms persist, get medical help.

6-12. DIARRHEA

Diarrhea can be treated by drinking tea leaves with water.

6-13. HEAT CRAMPS

Heat cramps are caused by an excessive loss of salt in the body. This loss of salt causes muscles to spasm, and they are unable to relax. After prolonged physical activity in a cold-wet environment, the signs and symptoms of heat cramps could appear in the arms, legs, and/or stomach. It is possible to experience heat cramps in cold-weather environment when undergoing extreme physical exertion.

SYMPTOMS

The soldier experiences muscle cramps in his arms, legs, and/or stomach. He is pale and has wet skin and is experiencing dizziness and extreme thirst.

TREATMENT

Have the soldier drink one fourth of a canteen of water. Give the soldier one canteen of water with one packet of salt stirred in. Have him drink the canteen of salted water over a 30-minute period. If cramping of muscles continues, evacuate the soldier to the nearest medical treatment facility.

> **NOTE**
>
> **Do not give the saltwater to a soldier who has an upset stomach. Have him drink a canten of unsalted water.**

6-14. HEAT EXHAUSTION

Heat exhaustion is caused by dehydration and loss of body salt from extreme physical activity in a cold-wet environment. It is possible to experience heat exhaustion in a cold-weather environment when undergoing extreme physical exertion.

SYMPTOMS

The soldier may feel dizzy, weak, and/or faint. The skin feels cool and moist to the touch. He may feel nauseated or have a headache.

TREATMENT

The following is a recommended treatment for heat exhaustion:

- Have the soldier drink one canteen of water to relieve the symptoms.
- If he complains of cramps, give him one canteen of water with one packet of salt stirred in. Have him drink the salt solution over a 30-minute period.

> **NOTE**
>
> **Do not administer salt if the soldier has an upset stomach. Have him drink a canteen of unsalted water.**

- Loosen all tight-fitting clothes.
- Elevate the feet above the level of the heart.
- If the soldier is unable to drink water due to an upset stomach or if the symptoms have not improved within 20 minutes, evacuate the soldier to the nearest medical treatment facility.

CHAPTER 7

Bivouac Routine

"Had we lived, I should have a tale to tell of the hardihood, the endurance and the courage of my companions which would have stirred the hearts of every Englishman. These rough notes and our dead bodies must tell the story."

CPT Robert Scott

7-1. GENERAL

The essential requirements for survival and mission accomplishment in the cold/arctic environment are warmth, food, and shelter. As covered in Chapter 3, you are provided with tents that can be erected quickly. Tents can be lost or destroyed and separated from the ahkio groups in the confusion of combat. When this occurs, timber and snow, which are usually available, will be used for shelters. A detailed explanation of the bivouac routine should be contained in the unit SOP.

7-2. BIVOUAC SITE

Bivouac sites, other than those that are permanent nature in rear areas, either are tactical or are designated to provide living accommodations within a defended position. The tactical bivouac is usually not occupied for more than one night, is established at or after last light, and needs little preparation. It is organized for a rapid and disciplined response to enemy action. The defensive position is likely to be occupied for longer periods and must be established accordingly. The layout and placement of tents and shelters must be coordinated within the defense and in accordance with the unit SOP.

7-3. SITE SELECTION

The selection of bivouac areas is important and needs careful consideration. The challenge of area selection varies with the tactical situation, weather conditions, and terrain. The bivouac area will be tactically located in accordance with the principles of security and defense.

There must be concealment from air and ground observation and wind. Forested areas pose few challenges in comparison to areas that are located above the tree line. In selecting areas, ensure that local camouflage materials are available for use with the snow camouflage nets. Areas that offer better bivouac sites than others are explained below.

FORESTED AREAS

Forests provide excellent sites. Camouflage is easy and building materials, firewood, and wood for insulation are readily available. Although forests provide the best bivouac site, temperatures there will normally be colder than those in open areas.

MARSHY GROUND

In winter, good sites are frozen swampy areas, the banks of streams, and the shores of lakes. Some swampy areas may not freeze because of warm springs or marsh gases. Deep valleys should be avoided since colder air gathers there.

OPEN AREAS

Avoid open areas because of strong winds, drifting snow, and poor concealment. If avoidance is not possible, then pitch tents where they can be sheltered by natural windbreaks. These are normally found in depressions or behind ridges of ice on lakes. In areas where natural wind falls do not exist, build snow walls to provide wind protection. In areas with high winds, snow gathers on the lee side, making it necessary to clear the side and tops of the tents, often to prevent the weight of the drifting snow from collapsing the tent.

MOUNTAINOUS AREAS

Mountainous terrain is characterized by strong winds, cold, and a general lack of concealment above the timberline. The wind over head creates extensive lees near cliffs and steep rock faces. Avalanches can occur on steep slopes; avoid their likely path. Signs to look for are:

- Heavy snow overlaying a crust of ice (usually the result of a freeze at the end of a period of thaw).
- New snow on steep slopes if the temperature is rising.
- Evidence of avalanches in previous years; an area cut through the trees (branches stripped off, and/or broken tree stumps at different heights) or a smooth clear strip of ground down the mountainside.

Figure 7-1. Location of tent.

Figure 7-1. Location of tent (continued).

When contact with the enemy is likely, the unit SOP will identify specific considerations when selecting the site for a tactical bivouac. For example:

- The location of the enemy.
- Enemy patrol habits.
- The air threat.
- Dominating features that identify enemy avenues of approach.
- The wind direction that can carry sounds of movement (the bivouac site should be downwind of likely enemy avenues of approach, especially when wind speeds are low).
- Availability of natural camouflage.
- Amount of moonlight and weather conditions. In good weather or bright moonlight, the bivouac site must be in an area of deep shadow (in the middle of the woods or on the northern slope of a ridge) or the tents will be silhouetted.

7-4. RECONNAISSANCE OF BIVOUAC SITE

The main body should be preceded by a reconnaissance party whose task is to lay out the bivouac site. They should:

- Identify the bivouac site and a false position.
- Establish a trail plan.
- Select defensive positions.
- Mark the location of tent sites.
- Identify areas for cutting boughs.
- Act as guides for the main body on its arrival.
- Provide security during occupation by the main body.

7-5. TRAIL PLAN

Trail plans must be established before entering the bivouac site. Normally, the incoming trail will be extended well beyond the area of the bivouac and may lead to a false position. Leading off the incoming trail establish another trail, usually at an extreme angle to entry. The trail network that leads to each subordinate unit area, and within, must follow the inside of the tree line. This prevents enemy observation of trails from the air or ground.

7-6. OCCUPATION

Guides from the reconnaissance party will arrange to meet the main body and identify the layout of the bivouac site. Once the unit is in position, the unit should:

- Identify a temporary area for weapons and equipment during establishment of the site.
- Identify the exact location and type of defensive positions to be built.
- Determine exact tent sites regarding defensive positions, natural shelter, and camouflage.
- Identify sleeping areas and type of shelters to be built when tents are not used.

7-7. BOUGH-CUTTING AREAS

Bough-cutting areas for bedding or improvised shelters should be for all personnel in the unit. These areas must be where activity is not likely — well behind the unit's defensive area or in a decoy bivouac site.

Cut boughs from the bottom portion of trees above the snow and leave the trees standing. Be sure to cut only a few boughs from each tree. A tree stripped of its limbs is easily observed from the ground and from the air.

Cut firewood near the bough-cutting area to allow the same trail to be used. Dead spruce, pine, or squaw wood makes the best firewood. Standing deadwood burns the best and hardwoods the longest. In an arctic and subarctic environment, true hardwoods are rare. Of the trees available, birch, tamarack, and spruce are the hardest woods. Burn the upper parts of dead trees during daylight because they give off lighter colored smoke. The lower parts of the trees contain resin and tar, which burn well but make more and darker smoke. In the daylight this will allow your position to be easily identified by air and ground observation.

7-8. GARBAGE

All garbage should be double bagged and transported to the rear or disposed of in accordance with the unit SOP. Units should never leave evidence of their presence. Carry all trash, if necessary, until it can be disposed of.

7-9. LATRINE

Normally there will be a central latrine within the bivouac site, if the site is not too widely spread. One latrine can usually serve the needs of a platoon-size element. It must be placed downwind of the bivouac site, but not so far that soldiers won't use it. The area identified can be as elaborate as a pit or cross-cut-type latrine that is made windproof by branches, snowblocks, and ponchos. It should be properly camouflaged. It can also be an area where plastic bags are located. Double-bag, tie, and transport all waste to the rear for disposal as identified in the unit SOP. Site urinal areas around a "dedicated tree" or the gasoline tripod, which will hide the discolored snow and will be easily spread about when leaving the bivouac site.

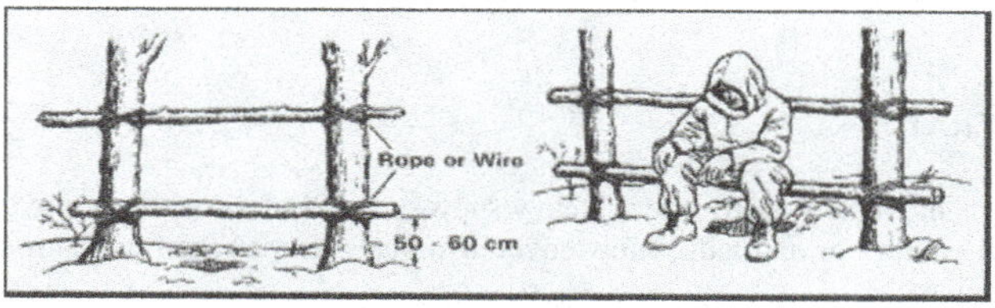

Figure 7-2. Cross-tree type latrine.

7-10. WATER POINTS AND SNOW AREAS

When obtaining water for drinking purposes by melting snow, set aside and restrict an area for this purpose only. The site must be upwind from the bivouac area and isolated from the latrine and garbage areas. The snow must be clean, white, and free of fungus from trees. NEVER use snow that is discolored. Remember — disinfect the water using purification tablets and bring it to a rolling boil for a minimum of 15 minutes.

7-11. STORAGE

Storage problems in the winter are increased by snow, low temperatures, thaws, and limited storage space. Space in any shelter is limited. Store only those items inside the tent that are affected by the cold, or must be immediately available. The remaining equipment should be centralized, well marked, and covered. Store ammunition and fuel separately. Place wood or metal containers on boughs or poles to keep the containers from freezing to the ground. The unit SOP will specify what equipment is placed inside or outside of the tent and where.

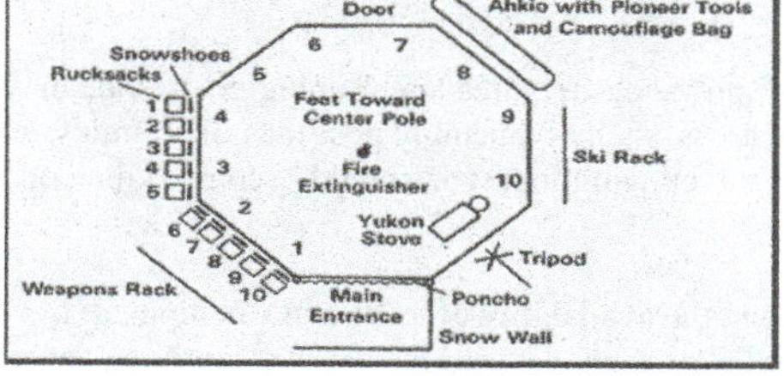

Figure 7-3. Ten-man tent.

CHAPTER 8

Individual Movement

"Anticipate every situation which can arise in battle and think out your reaction to it well in ADVANCE."

Hoote

8-1. GENERAL

Snow affects movement many ways. Heavy snow cover hinders or slows movement, either cross-country or on roads. Snow cover also blankets many terrain features and hides obstacles.

The effect of snow on movement varies greatly with its depth and physical characteristics. Men on foot cannot travel well over flat terrain or roads when the snow is deeper than 30 centimeters. However, movement can be improved by the use of skis or snowshoes.

SKIS	SNOWSHOES
• Faster • Less Work • Harder to Learn	• Slower • More Work • Easy to Learn

Figure 8-1. Using skis or snowshoes.

8-2. SKIS

Skis provide greater speed in moving, particularly over cleared trails. Troops mounted on skis and towed behind small unit support vehicles, or similar vehicles (ski-joring), are effective means for rapid cross-country movement. In heavy vegetation, skiing will always be more physically taxing than snowshoeing.

8-3. SNOWSHOES

Learning how to use snowshoes requires less training. Snowshoe movement is more practical in confined areas, such as assembly positions, field trains, and mortar and artillery positions. If traveling on foot, you should become skilled on both skis and snowshoes.

Your ability to use, and the availability of, over-snow equipment is a major factor in learning to travel well over snow-covered terrain. With previous training in cross-country skiing and if equipment is available, travel on skis is recommended. In most snow conditions and over most types of terrain, skis provide the fastest and most energy-saving form of travel. Use of snowshoes requires little training, but your speed will be slower and travel more exhausting.

Snowshoes are **more** exhausting than skis. However, a light crust on the surface of the snow prevents skis from sinking and provides for fast and easy skiing. A crust hard enough to support a man makes travel on foot feasible, but even then travel on skis is recommended.

In a survival situation, improvise equipment for travel if snow is loose and deep. Make snowshoes of willow or other springy wood using a wood separator and thong, wire, cord, etc.

USE SKIS WHEN:

- Speed is essential.
- Long distance must be traveled.
- Stealth is necessary.
- Conditions allow.

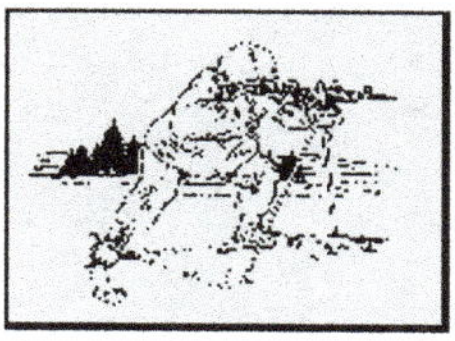

USE SNOW SHOES WHEN:

- Moving through heavy brush.
- Speed is not essential.
- Troops are not proficient on skis.

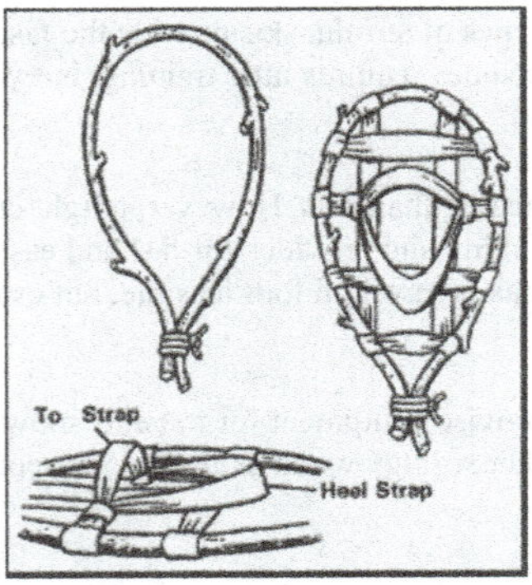

Figure 8-2. Improvised snowshoes.

CHAPTER 9

Land Navigation

"No - I've never been lost; but for about three weeks I was kinda puzzled."

Brigadier General Fergusson

9-1. GENERAL

In some cold-weather environments, accurate up-to-date maps may not be available and landmarks may be few or nonexistent. When the terrain is snow covered, one piece of ground looks like the other. Whatever the circumstances, you must be able to find your way — if not by map reading, by some other means of navigation.

9-2. CHALLENGE OF LAND NAVIGATION

Only poor road nets may exist in some areas. These will be of little use in cross-country navigation. If large-scale maps exist, they may be unreliable.

Long nights, falling or blowing snow, and fog limit visibility in winter. Tracks and landmarks may be erased by heavy snowfall. Accurately identify landmarks, through both ground and air observation, to prevent mistakes.

Conditions of daylight are different from those found in temperate regions (moderate climate). In winter there are long nights and short days.

Magnetic disturbances may be common, thereby making compass readings unreliable. Magnetic declinations may be extreme and can vary considerably at close points.

Many small lakes, ponds, and creeks will be found that are not on the map. The size and location of waterways may vary from year to year and from season to season and cannot be relied on to pinpoint a position. Large lakes and rivers can be used as important aids to navigation and movement.

9-4. DISTANCE AND DIRECTION

When traveling from one point to another, it is necessary for you to know the distance between them beforehand, then plot your journey accordingly.

9-5. MEASURING DISTANCES

PACING

In temperate zones, pacing is the simplest method of measuring distance. In cold-weather areas, snow cover and the use of skis and snowshoes make pacing challenging, and it must be practiced repeatedly.

USE OF FIELD WIRE

The best method for accurate ground measurement is for two men to use a piece of salvage field wire 50 meters long. The lead man marches off in the desired direction trailing the wire behind him. The second man signals by jerking the wire when the first has gone 50 meters and then moves forward to the first man's position. The first man marks the spot and continues forward for another 50 meters. This is repeated until the desired distance is reached.

ESTIMATION

If no other method of measurement is available, the length of time spent in steady progress along a route may be used to estimate the distance traveled. The accuracy of this method depends on how well you know your marching speed over all types of terrain on foot, skis, or snowshoes.

9-6. DETERMINING DIRECTION

COMPASS

The use of the compass is the most common way to determine direction. The magnetic field becomes weaker as you get nearer to the magnetic poles and can be easily affected by other objects. It is important to make sure that the compass is not affected by iron fragments in the ground, vehicles, or metal objects on you.

THE SUN

There are two ways to determine direction by the sun:

- At noon, in the northern hemisphere, shadows fall to the north of objects; in the southern hemisphere, to the south of objects.
- Estimate direction by using a watch and the sun. In northern latitudes the hour hand is pointed toward the sun. An imaginary line halfway between the hour hand and 1200 will point south. In southern latitudes the procedure is different. Then point the 1200 toward the sun. An imaginary line halfway between the hour hand and 1200 will point north.

NATURE

Nature will help in determining direction as in the following samples:

- Single evergreen trees, growing away from other trees, will always be more bushy on the south side.
- The bark of birch and poplar trees is whitest on the south side and darkest on the north.
- You should know from which direction the prevailing winds blow. Trees will be bent in the direction the wind is blowing or expected to blow.

THE STARS

The North Star may be used to determine true north. This star can be easily found by its position relative to the Big Dipper. In latitudes under 70 degrees when traveling north, the North Star makes a good steering mark since its bearing is usually only 1 degree from true north and is never more than 2 1/2 degrees away. In higher latitudes above 70 degrees, the North Star is too high in the sky to indicate good direction.

A star near the horizon can be used with a bearing within 2 degrees of your compass course. However, it will only be good for up to 30 minutes when you are traveling north and only 15 minutes when traveling south. After these intervals, select another star.

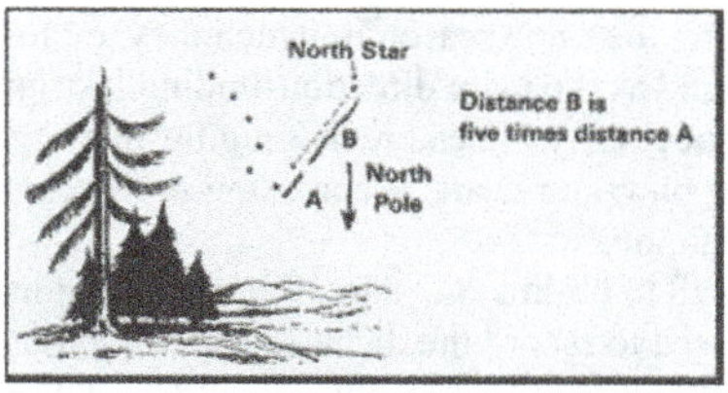

Figure 9-1. Determining direction using North Star.

South of the equator, the constellation Southern Cross will help you locate the general direction of south and any other direction. This group of four bright stars is shaped like a cross that is tilted to one side. The two stars forming the long axis, or stem, of the cross are called pointers. From the foot of the cross, extend the stem five times its length to an imaginary point. This point is in the general direction of south. From this point, look straight down to the horizon and select a steering point.

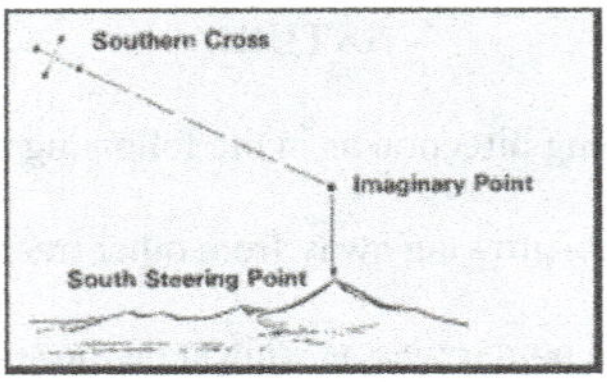

Figure 9-2. Determining direction using Southern Cross.

9-7. NAVIGATION BY DEAD RECKONING

Dead reckoning consists of plotting and recording a series of courses from a known point, each measured for distance and direction. These courses lead from the starting point to the destination. This lets the navigator determine his position at any time, either by following his plot or by comparing his actual position on the ground to his plotted course. All the details are previously estimated and recorded.

EQUIPMENT FOR DEAD RECKONING

- Maps are used for choosing the route and for plotting the route taken as the march progresses.
- Aerial photographs are usually used in addition to maps. They must be vertical photos taken during the same season in which they are to be used.
- The compass is used as the main direction-finding instrument.
- The protractor is necessary in dead reckoning for plotting direction and distance on the map, aerial photo, or chart. When using it, remember to convert magnetic azimuth to grid azimuth.
- A route card is used to outline the plan of the proposed march. Keeping a log is mandatory. It is used to record the distance and direction.
- The log is a written record of where the soldier has started, the distance and direction traveled, and his position at any time in relation to the planned route, destination, and starting point.

THE DRILL OF DEAD RECKONING

Know the starting point and destination and, having examined the map or aerial photo for the best route, prepare a route card. This describes each leg of the proposed route in terms of distance and direction. As the march continues, keep a careful record of each bearing taken and the distance, and record this in the log. There may be a change to the planned route because of terrain or enemy action. Therefore, the necessary changes are made using the map or aerial photo as a guide, and they are recorded in the log.

9-8. HINTS FOR THE NAVIGATOR

The navigator must keep the compass warm to speed up recording his position. When no steering marks exist to the front, he should march on a back azimuth. A steering mark may be a natural or man-made structure to the rear. If the steering mark is not clear, he must keep an eye on it after taking an azimuth so that it doesn't become lost. Never take magnetic azimuths in the area of metal objects.

When visibility is poor, only close-in steering marks will be available. Under these conditions try to pick up steering marks along the correct azimuth. This can be done correctly only when the route to each steering mark follows a straight line. Take compass checks often to ensure that the correct azimuth is being followed. Set the compass for a night course. A steady unshifting wind will also aid in keeping direction.

Depending on the area of the world, other signs can be used. For example, in the barren and semibarren lands of eastern Canada, snowdrifts point to the northwest and windblown trees to the southeast because of prevailing winds. In Scandinavia, scratches on rocks and cliffs caused by the glacial period (ice age) run northwest to southeast.

CHAPTER 10

Survival

"Each day the number of my dogs dwindled and the sleds were broken up to cook those animals that we ate ourselves."

Peary, Arctic Expedition, 1906

10-1. GENERAL

Your mission may take you to an arctic or subarctic region. You expect to have all personal equipment and other members of your unit with you wherever you go. But you could find yourself alone in a remote area with little or no personal gear.

10-2. ACTION WHEN LOST

Brief the soldier on the route before setting out on a march; he should memorize details if possible. A soldier can become separated from the main body of troops, especially if an action is fought. This applies mainly to night battle or bad weather. It can also happen if he is careless and march discipline is poor.

WHEN LOST WITHIN A KNOWN LOCALITY

If all is quiet and there are no battle sounds or aircraft to guide you to friendly lines, go back to the last known position. If you cannot go back, consider carefully the journey already made and the distances and directions from the starting point.

Opinions should be taken from the group. Search parties should be sent out to find the next known steering mark. Trails must be marked carefully so that the search party can return to guide the main group forward or to rejoin the group, should their search be unproductive. Meanwhile, the rest of the group should seek shelter.

If the search party fails to find any known location, perform the steps below, but remember to think clearly and keep calm.

CONDUCT WHEN LOST

When you are certain that you are lost:

- Stay together. Meet your emergency as a group; don't wander off as individuals; never allow stragglers.
- Take shelter. If you have a tent, set it up and get in it. If you do not have a tent, build one of the improvised shelters. The type of shelter is not important as long as you protect yourself from the weather.

- Keep warm. If you are lost in a forest, light a fire (depending on the enemy situation). If you are lost in an area where there are no trees, do not use your gasoline stove to heat your shelter. Use your stove only to prepare hot food and drinks. Conserve your fuel supply; you won't get any more. Get into your sleeping bag to stay warm.
- Check food supply. All the food and water in the group should be collected and rationed systematically. Supplement the food supply by shooting large animals and trapping or snaring small ones.
- Check communications. If you have a radio set, use it to contact your parent unit for help. If you are in a forest, prepare fires to attract search planes. Smoke is better by day and flame at night. Mirrors, ground signals, etc., are helpful in attracting search parties. Arrange necessary ground/air signals for search planes.
- Rest. The more you rest the less energy you expend and the less food you need.
- Adhere to survival tips. Carry safety matches inside the butt plate of your rifle or in your pockets. Carry candles, which are easy to light, provide heat and light, and help in starting a fire. If you have no matches, fire may be produced by striking the flint on your waterproof match box with the blade of a knife and collecting sparks in dry tinder. Dry moss and bark make good tinder to start your fire. Fires should be built in layers with sticks being laid out like the spokes of a wheel.
- Carry a two-cell flashlight. It can be seen from the air or on the ground as far as line-of-sight will permit.
- Ground/air signals can be made with small fires to form the identification characters (see Appendix C).
- Survival depends on clear thinking and resourcefulness. The main things to remember are to keep calm, think, try to help each other, keep together, and keep warm.

10-3. IMPROVISED SHELTERS

Sometimes tents or other regular shelters are not available. In summer, if the weather is mild, you may need protection only from insects. In the winter, however, you cannot stay in the open for long periods unless you are moving. In case of emergency, you must know how to protect yourself from the effects of the weather.

If natural shelters such as caves or rock ledges are available, they should be used. If they are not, a temporary improvised shelter must be built.

The type of shelter to be built depends on the equipment and materials available. By the proper use of materials available, some sort of shelter can be built during any season of the year. In open terrain a shelter can be built using ponchos, canvas, snow blocks, or other materials. Snow caves, snow trenches, snow houses, or snow holes may be built in the winter. In the woods, a lean-to is normally preferred to other types of shelter. In cold-weather areas, nature provides you with the means to prepare a shelter. Your comfort, however, greatly depends on your initiative and skill at improvising.

PONCHO SHELTERS

The simplest type of shelter can be made by pulling the poncho over your sleeping bag. For additional comfort, various types of shelters and lean-tos can be made by attaching your poncho to trees, tree branches, or poles.

To build a one-man shelter from one poncho, spread the poncho on the ground, hood side up. The hood opening must be tightly closed by adjusting and tying the drawstrings. The poncho is raised in the middle of its short dimension to form a ridge, with the stake out at the corners and sides. Snow, sod, or branches are used to seal the side and one end of the shelter to provide additional protection from the wind and to retain heat.

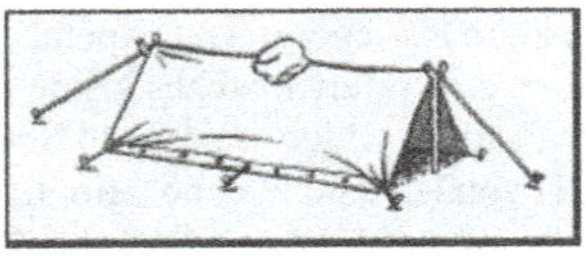

Figure 10-1. One-man shelter.

A two man shelter can be made from two ponchos. Spread the ponchos on the ground with the hood side up and the long sides together so that the snap fastener studs of one poncho may be snapped into those of the other poncho, The ponchos are raised where they are joined to form a ridge. The shelter is then staked out at the sides and corners. A third poncho may be snapped into the other ponchos to form a ground cloth.

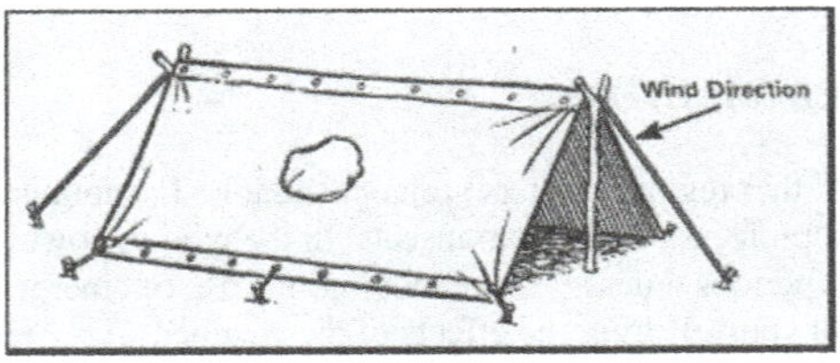

Figure 10-2. Two-man shelter.

LEAN-TO SHELTER

The lean-to shelter is built in forested areas and is made of trees and tree limbs. A poncho, a piece of canvas, or a parachute, in addition to boughs, may be used for covering. Depending on the number of people to be sheltered, two types of lean-tos (single and double) are built. Boughs should be laid in the shelter like overlapping shingles (the same as on a live tree).

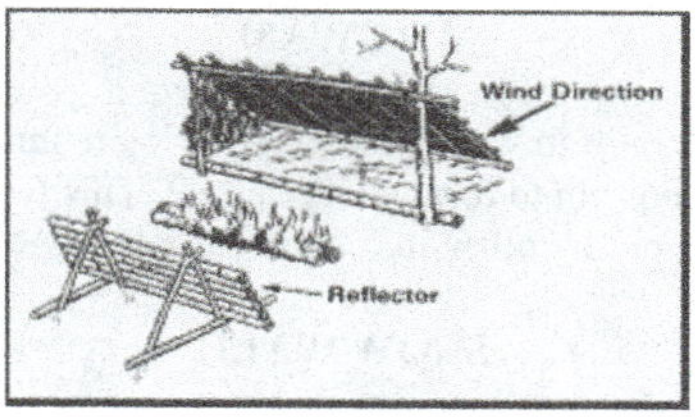

Figure 10-3. Single lean-to.

To save time and energy, use two trees the right distance apart and sturdy enough to support the crosspiece. If suitable trees are not available, use two forked poles or build two A-frames to hold the crosspiece. A large log is laid at the back for added height. Stringers about 3 meters long and 10 centimeters in diameter are then placed 45 centimeters apart from the crosspiece over the top of the log in the rear. Available coverings, ponchos, boughs, and so forth are then placed on top of the stringers. A double lean-to is made by building two single lean-tos facing each other with the fire in between.

Any kind of an open fire can be used to heat the lean-to. The best kind for a large lean-to is the log fire. Precautions must be taken to prevent the fire from burning too hot and burning down the shelter. A fire reflector may be built of green logs or poles to reflect the heat into the shelter and to serve as a windbreak to keep the fire burning steadily.

TREE SHELTERS

In wooded areas, the tree-pit shelter furnishes temporary protection. Select a large tree with thick lower branches and surrounded with deep snow. Enlarge the natural pit around the tree. The walls and floor are lined with branches and boughs. Try not to disturb the snow that may be on the branches.

An emergency shelter for one man can be built by cutting down a coniferous tree at a point about 1 meter from the ground.

Figure 10-4. Tree-pit shelter.

After the tree has fallen, the branches on the underside are trimmed away and the cut material used for insulation on the ground.

WIGWAM

A wigwam or tepee can be built in wooded areas by tying a number of poles near the top and spreading them at the bottom to form a large circle. This framework is then covered with ponchos, a parachute, or any other suitable material.

SNOW WALL

In open terrain with snow and ice, a snow wall may be built for protection from strong winds. Blocks of compact snow and ice are used to build a windbreak. Be careful not to get covered up by drifting snow, which may accumulate on the lee side of the wall.

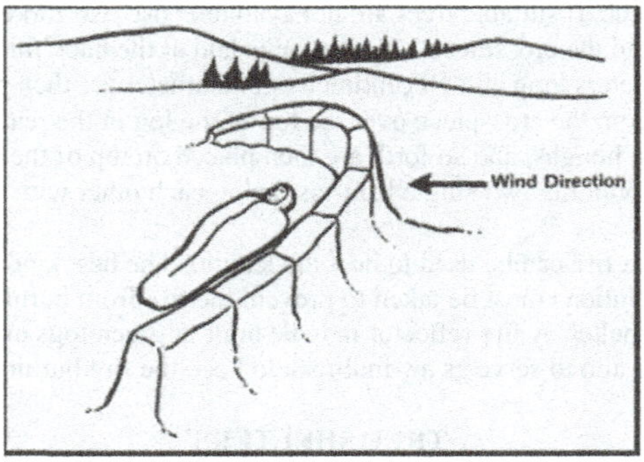

Figure 10-5. Snow wall.

SNOW TRENCH

A snow trench provides shelter quickly. It is built by burrowing into a snowdrift or by digging a trench in the snow and making a roof of ponchos or ice and snow blocks supported by skis, ski poles, or snowshoes. Boughs can be used for covering the roof and for the bed. Figure 10-6 shows one method of roofing over the trench by leaning snow blocks against each other.

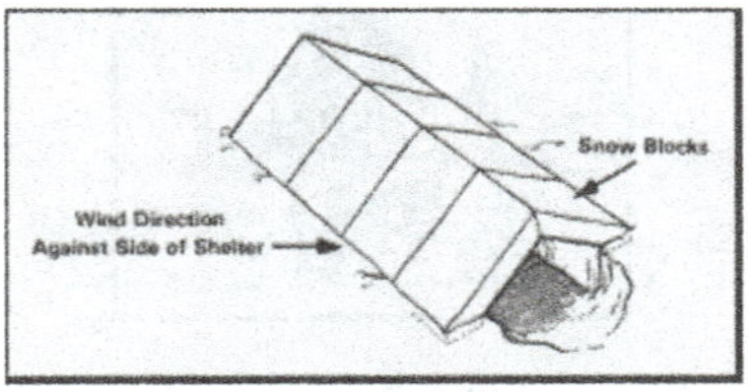

Figure 10-6. Snow trench.

SNOW CAVES

A snow cave can be used as an improvised shelter in the open areas where deep and compacted snow is available. Normally, a suitable site can be located on the lee side of a steep ridge or riverbank where drifted snow collects, and it is wind packed.

The tunnel entrance must give access to the lowest level of the chamber. This is the bottom of the pit, where cooking is done and equipment is stored. The snow cave must be high enough to provide comfortable sitting space. The sleeping area must be on a higher level than the highest point of the tunnel entrance, where there is warmer air.

The roof must be arched for strength so that drops of water forming on the inside will not fall on the floor. This water will follow along the curvature of the roof and sides, glazing over the walls when frozen. The roof must be at least 30 centimeters thick. The size of the snow cave depends upon the number of men that will occupy it. A large cave is usually warmer and more practical to build and maintain than several small caves.

The shape of the snow cave can be varied to suit conditions. When the main cave is built, short side tunnels are dug to make one- or two-man sleeping rooms, storage space, latrine, or kitchen space.

In building the snow cave, locate a deep snowdrift at least 3 meters deep. Newly fallen, powdery, or loose snow should not be used. The depth of a snowdrift may be tested with a sharpened branch or young tree about 4 meters long, or with the ski or shorter ski pole. The entrance must be chosen carefully so that wind will not blow into the cave or block the entrance by drifting snow. The entrance should be built so that it is about 45 degrees from the downwind side. A small tunnel is burrowed directly into the side of the snowdrift for 1 meter. A chamber is excavated from this tunnel. Excavation is done to the right and left, so that the length of the chamber is at right angles to the tunnel entrance. Personnel doing the digging will become wet from perspiration and from the snow inside the cave. They should wear the minimum amount of clothing to ensure that they have a change of dry clothing when finished.

The cave can be heated with the one-burner gasoline stove or with candles. The fires should be put out when personnel inside the cave are sleeping, thus reducing the danger of fire, asphyxiation, or carbon monoxide poisoning. If the weather is severe and it is necessary to have a fire while sleeping, an alert fire guard must be posted in each cave. The ventilation holes must be inspected every 2 or 3 hours to ensure that they have not become clogged by snow or ice.

To ensure that the cave is warm, the entrance should be blocked with a rucksack, poncho, or snow block when not in use. All available material such as ponchos, cardboard from ration boxes, brush, and boughs should be used for ground insulation.

Walking on the roof may cause it to collapse. At least two ventilator holes are used, one in the roof and one in the door. A ski pole can be stuck through the roof ventilator to clear it from the inside. Extra care must be taken to keep air in the cave fresh when cooking or heating. The entrance should be marked by placing a pair of skis, or other equipment, upright on each side of the entry way.

Building a snow cave requires practice. Large shelters are usually more efficient for squad-size units and larger units. The most critical decision is when and where to stop and build the shelter.

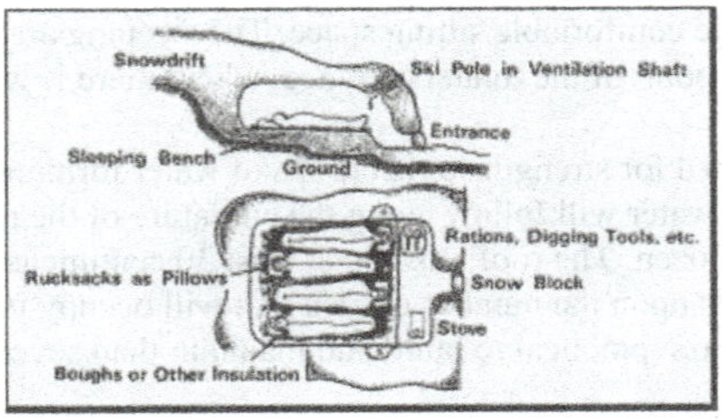

Figure 10-7. Snow cave.

INDIVIDUAL SNOW HOUSE

If you are alone and have need of a shelter quickly, you can build a small individual snow house. It is box-shaped, and the only building materials needed are snow blocks. Cut snow blocks 40 by 40 by 80 centimeters. Lay them on edge to a length of 2.5 meters and a width of 1 meter. Build the walls up to a height of about 1 meter. When you have the walls built, make an A-shaped roof by leaning together pairs of blocks. Fill in the ends with snow blocks and chink all the crevices with soft snow. Cut a hole in the end of the leeward side large enough to crawl into. When you are inside, make one or two small holes near the top for ventilation. Block up the end with a snow block or with your rucksack.

Figure 10-8. Individual snow house.

SNOW HOUSE OR IGLOO

An igloo makes an excellent shelter in snow and glacier areas when other materials are not available or when the snow is so hard that it can be cut into blocks. Cracks in the igloo are then covered with loose snow. To increase the head room and to prevent water droplets from forming inside the igloo, the corners of the blocks may be cut off to form a round and smooth interior.

A shelter may be built over the entrance in the same manner as the igloo, and it should be 1 to 2 meters long and about 1 meter wide and 1 meter high. This shelter provides more protection and warmth inside the igloo and a place to store equipment. Blocks of snow or other material may be placed at both the outside and the inside entrances of the tunnel for more warmth. Make a ventilation hole near the top and at the same height as the fifth or sixth row of snow blocks. The hole can be kept open by placing a ski pole or long stick into it and clearing it out as needed. The igloo is heated with a one-burner stove or with candles.

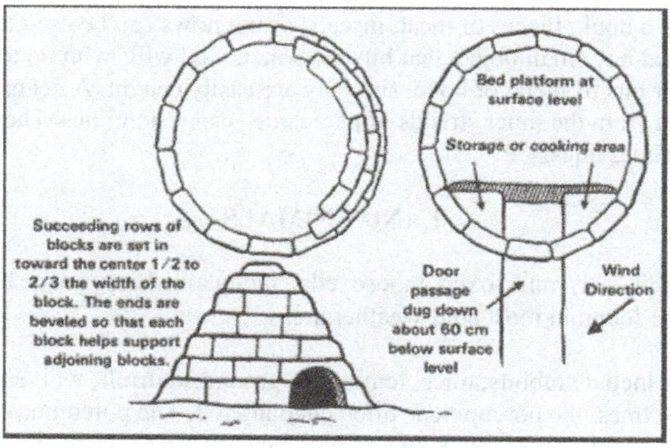

Figure 10-9. Snow house.

10-4. FOOD

The chances of finding different types of food vary in cold-weather areas depending on the time of year and the place. Shores normally are scraped clean of all animals and plants by winter ice. Food can still be found north of the timberline, even when such foods as mice, fish, and grubs are not available.

STORAGE AND PRESERVATION

If a large animal is killed or a large number of small game are found, you should store or preserve some of the meat for future use. Freezing is the best way to preserve fresh meat or fish. It may be necessary to protect your supplies from small animals. This can be done by hanging these supplies about 2 meters from the ground.

FISH

There are few poisonous fish in cold-weather areas, but some fish, like the sculpin, lay poisonous eggs. The black mussel may be poisonous at any season and is as deadly as strychnine. In cold regions do not eat shark meat; it is poisonous. In coastal streams and rivers, salmon moving upstream to spawn may be plentiful; however, their flesh deteriorates as they travel away from the coast, making them unfit to eat except in an emergency. A salmon that has been in fresh water long enough to turn a brilliant red is decaying and may be unfit to eat.

Coastal waters are rich in all seafood. Grayling, trout, white fish, burbot, and pike are common to the lakes, ponds, and arctic coastal plains of North America and Asia. River snails or freshwater periwinkles are plentiful in the rivers, streams, and lakes.

Fish can be speared, shot, netted, hooked, caught by hand, or stunned by a rock or club. Fish hooks can be improvised from pins, pieces of wire, or any other metal object that can be bent into a hook. Pieces of meat, insects, or minnows can be used for bait. Some fish will nibble at any small object that hits the water. Cod will swim up to investigate strips of cloth or bits of metal or bone, and they are easily caught. A net can be made out of stout twine or from the inner strands of parachute suspension lines. These strands can also serve as fishing lines.

LAND ANIMALS

Deer, caribou, reindeer, musk-oxen, moose, elks, mountain sheep, goats, bears, and other such animals are found in most cold-weather areas.

Tundra animals include rabbits, mice, lemmings, ground squirrels, wolves, and foxes. Where there are trees, the porcupine is often encountered. The porcupine can be easily clubbed or shaken out of trees. Pick up a porcupine only after it is dead.

Hunting is better during the early morning or late evening when the animals are moving to and from feeding and bedding grounds and water. Large game such as moose and caribou should be shot just back of the front shoulder or, if the range is short and you are sure of your shot, in the neck. These large animals are fairly easy to stalk and kill, and their skins are also useful for survival. Bears should be shot just forward of the front shoulder to stop them or, if you are sure of your shot, in the neck. To successfully hunt these land animals you should know something of their characteristics:

- Caribou or reindeer are usually very curious. It is possible to attract them near enough for a shot by waving a cloth and moving slowly toward them on hands and knees.
- The technique of imitating a four-legged animal may also cause a wolf to come closer to a hunter.
- Moose may be found in heavy brush or around the shoreline of lakes.
- Mountain goats and sheep are cautious and hard to approach. They can be surprised, however, by moving quietly downwind while they are feeding. If possible, stay on higher ground than they are on.

Figure 10-10. Hunting land animals.

- Musk oxen leave cattle-like tracks and droppings. When alarmed, they group together in a circle with their heads facing out and remain in that position unless approached, then one or more bulls may charge.
- Bears are always dangerous. A wounded bear is most dangerous and should not be followed into cover. The polar bear is a tireless, clever hunter with good sight and an extraordinary sense of smell. Be careful that you don't become the hunted instead of the hunter.
- Rabbits often run in circles and return to the same place when they are frightened. A running rabbit can sometimes be stopped by whistling. Snares made of wire and placed on small game trails, arranged so that the animal will place his head through the loop, are an excellent means of catching rabbits.
- The quick-kill method can be used on most small animals and sometimes on larger animals after they have been shot and are on the ground. The method is simple and consists of cutting the animal's throat with a sharp knife. This system not only kills the animal but bleeds him at the same time.

Figure 10-11. Snaring a rabbit.

SEA ANIMALS

During winter and spring, sea mammals — seals, walruses, and polar bears — are found on the frozen pack ice and on floes in open water.

Seals are hard to approach but can be stalked. Keep downwind and avoid sudden moves. A white camouflage suit will help. Advance only when the animal is sleeping. Do not eat the liver of the bearded seal, because its high vitamin A content may cause you to be sick.

Walruses are found on pack ice and must be approached by boat. Because of its large size and boldness, the walrus is one of the most dangerous animals in the arctic.

Polar bears are found in almost all arctic regions but rarely appear on land. Avoid them if possible. If it is necessary to kill one for food, do not eat the liver. The liver is high in vitamin A, which may cause you to be sick. Never eat any bear meat unless it is cooked; it is always diseased.

BIRDS

The breeding ground of many birds is in the arctic. Ducks and geese build their nests near ponds on the coastal plains during the summer and provide an important source of food. Grouse and ptarmigan live in mountainous terrain and brush-covered areas in arctic and subarctic areas. Sea birds may be found on cliffs or small islands on the coast. Sea birds, ravens, and owls are useful for food.

In winter, owls, ravens, grouse, and ptarmigans are the only birds available. Ptarmigans and grouse are easily approached and are very tame. Although they are hard to locate because of the protective coloring, they provide a good source of food because they can be killed with a stone or a club.

PLANTFOOD

Most plants that grow in cold regions can be eaten. The water hemlock and the baneberry are the only poisonous plants. The water hemlock is one of the world's most poisonous plants. It can be recognized by where it grows, always in wet ground, and by the following characteristics: a hollow, partitioned bulb at the base of a hollow stem, spindle-shaped roots, and a foul odor, which is very noticeable in the root and bulb. The water hemlock is abundant in marshes near southern beaches and around marshy lakes in interior river valleys. It is never found on hillsides or dry ground. The baneberry is normally found on hillsides or in forested bog areas. Some mushrooms are poisonous and should not be eaten. Although the cold-region areas have many kinds of berries that are edible, most will not be available in the winter.

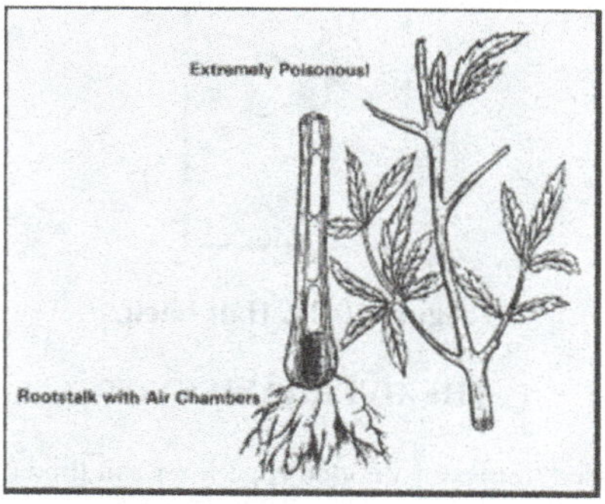

Figure 10-12. Water hemlock.

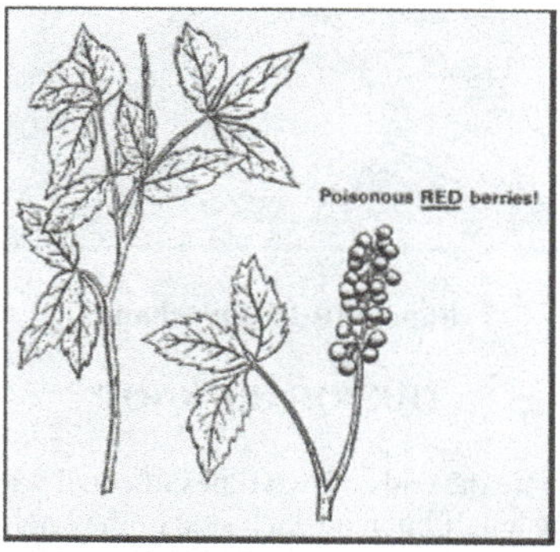

Figure 10-13. Baneberry.

10-5. USEFUL KNOTS

A rope is only a limp coil of hemp or nylon. It is useful only when you are able to attach it to itself or another object. The six basic knots described below will fit all your needs for tying rope together or onto equipment or personnel.

THE HALF HITCH

The half hitch is used to secure the end of a rope to some other object or to secure other knots.

Figure 10-14. Half hitch.

THE OVERHAND KNOT

The overhand knot is used to make a knotted rope for a handline or to temporarily whip the end of a rope to keep it from unraveling.

Figure 10-15. Overhand.

THE SQUARE KNOT

The square knot is used to tie the ends of two ropes of equal width together. The knot should be secured by tying a half hitch on each side of the knot. If properly tied, the square knot will not slip or jam. Be careful to follow Figure 10-16 or you may end up with a "granny knot," which can slip and jam.

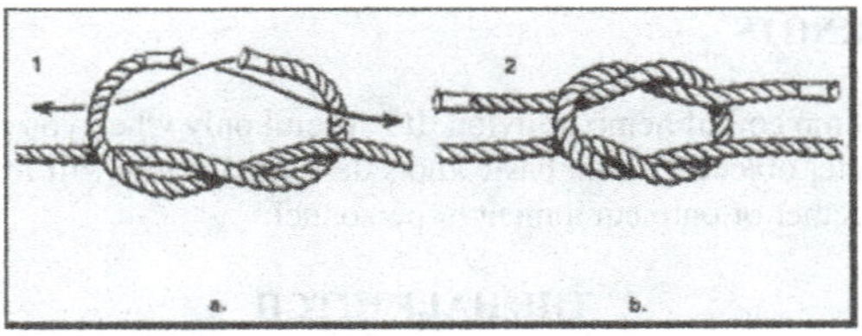

Figure 10-16. Square.

THE CLOVE HITCH

The clove hitch is the most useful of the hitches. It can be used to tie the end of a rope to a post or eye, or it can be used to secure the middle of a rope without using the ends.

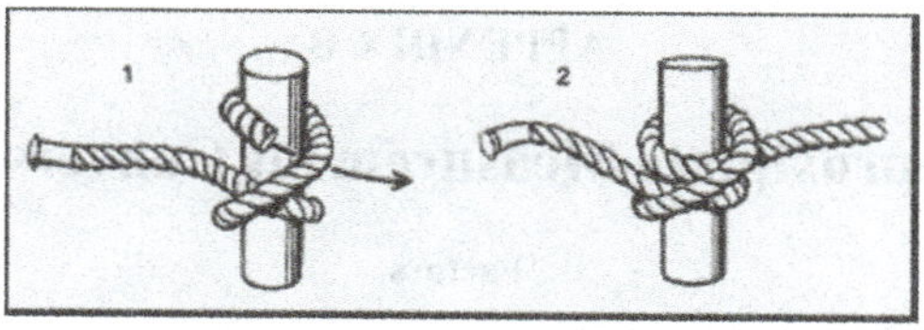

Figure 10-17. Clove hitch.

THE DOUBLE SHEET BEND

The double sheet bend is used for tying two ropes together of different width.

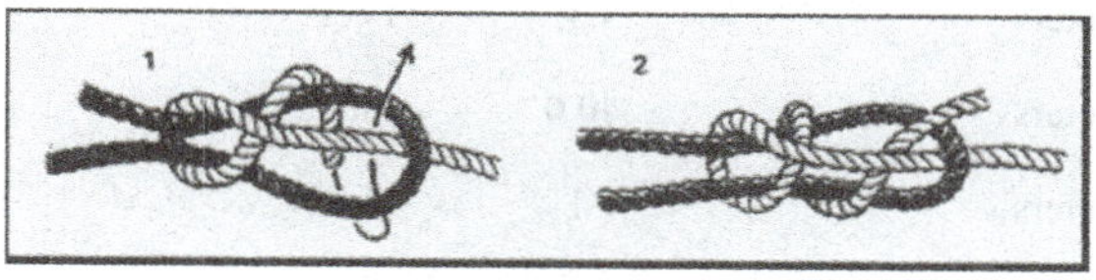

Figure 10-18. Double sheet bend.

THE BOWLINE

The bowline is used to form a loop on the end of the rope that will not slip.

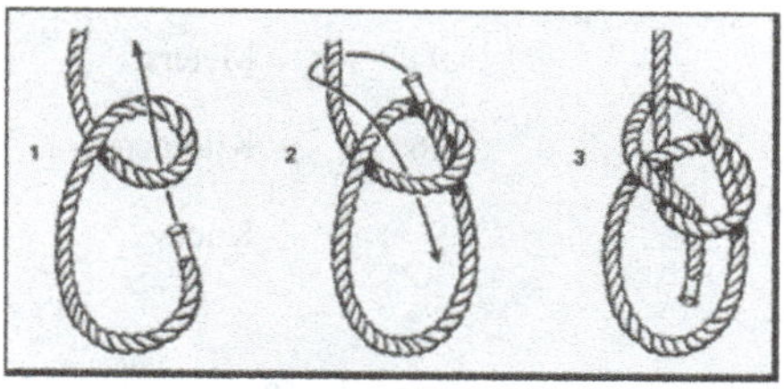

Figure 10-19. Bowline.

Remember the qualities of a good knot: it is easy to tie, it does not jam and become hard to untie, and it does not slip when weight is put on it.

APPENDIX B

Approximate Measurement Conversion

Factors

Multiply	*By*	*To Obtain*
Centimeters	.03	Feet
Centimeters	.4	Inches
Meters	3.3	Feet
Meters	40.0	Inches
Meters	1.1	Yards
Kilometers	.6	Miles
Knots	1.2	MPH
Inches	2.5	Centimeters
Feet	30.0	Centimeters
Yards	.9	Meters
Miles	1.6	Kilometers
MPH	.9	Knots

For ease in conversion the above factors have been rounded off.

References

Suggested Readings

DA Pam 20-201	Military Improvisations During the Russian Campaign
DA Pam 20-203	Russian Combat Methods in World War II
DA Pam 20-231	Combat in Russian Forests and Swamps
DA Pam 20-234	Operations of Encircled Forces: German Experiences in Russia
DA Pam 20-236	Night Combat
DA Pam 20-240	Rear Area Security in Russia: The Soviet Second Front Behind the German Lines
DA Pam 20-269	Small Unit Actions During the German Campaign in Russia
DA Pam 20-271	The German Northern Theater of Operations, 1940-1945
DA Pam 20-290	Terrain Factors in the Russian Campaign
DA Pam 20-292	Warfare in the Far North
FM 9-207	Operation and Maintenance of Ordnance Materiel in Cold Weather (0° to -65°F)
FM 31-70	Basic Cold Weather Manual
FM 31-71	Northern Operations
FM 90-6	Mountain Operations
FM 90-11 (TBP)	Winter Operations
TC 90-11-1	Military Skiing

"Perhaps the most valuable result of all education is the ability to make yourself do the things you have to do when it ought to be done, whether you like it or not."

Hosley

Glossary

ALICE	all-purpose lightweight individual carrying equipment
C	Celsius
CLP	cleaner lubricant preservative
ECWCS	extended cold-weather clothing system
F	Fahrenheit
FM	field manual
GP	general purpose
SFP	synthetic fiber pile
SOP	standing operating procedure
TBP	to be published
VB	vapor barrier